PROGRESS IN CHAOS AND COMPLEXITY RESEARCH

PROGRESS IN CHAOS AND COMPLEXITY RESEARCH

FRANCO F. ORSUCCI
AND
NICOLETTA SALA
EDITORS

Nova Science Publishers, Inc.
New York

LIBRARY OF CONGRESS CATALOGING-IN-PUBLICATION DATA
Available upon request

ISBN: 978-1-60456-375-7

Published by Nova Science Publishers, Inc. ✦ New York

CONTENTS

PREFACE

This book presents new leading-edge research on artificial life, cellular automata, chaos theory, cognition, complexity theory, synchronization, fractals, genetic algorithms, information systems, metaphors, neural networks, non-linear dynamics, parallel computation and synergetics. The unifying feature of this research is the tie to chaos and complexity.

Chapter 1 - The present work introduces an analysis framework for the detection of transient synchronized states in multivariate time series. In case of linear data these states exhibit a dramatic increase and subsequent decrease of time scale in the signal, whereas the corresponding instantaneous phases show transient mutual phase synchronization. We propose a single segmentation algorithm for both data types, which considers the space-time structure of the data. Applications to linear and phasic simulated signals illustrate the method. Further applications to event-related brain potentials obtained from an auditory oddball experiment while real car-driving reveal the lack of the cognitive component P300 in an experimental condition. Further the obtained results indicate attention effects in the event-related component N100 and shows dramatic latency jitters between single datasets. A comparison of the proposed method to a conventional index of mutual phase synchronization demonstrates the superiority of considering space-time data structures.

Chapter 2 - Assuming a mathematical model based on van der Pol oscillator, we simulated the time course of the latent and acclaimed phases of the psychiatric pathology called bipolar disorder. Results were compatible with the analysis of experimental time series data of mood variation previously published by Gottschalk A. et al., (1995). Furthermore we performed Recurrence Quantification Analysis (RQA) of time series data generated by our mathematical model and we found that the obtained values for Recurrences, Determinism and Entropy may be considered as indexes of the increasing severity and stage of the pathology. We consequently suggest that these variables can be used to characterize the severity of the pathology in its observed stage. On the basis of the model, an attempt has also been made to discuss some aspects of the complex dynamics of the pathology. Results suggest that stochastic processes in mood variation of normal subjects play an important role to prevent mood from oscillating in a too rhythmically recurrent and deterministic way, as it occurs in bipolar disorder.

Chapter 3 - We present an efficient parallel algorithm for the shortest path problem in weighted interval graphs for computing shortest paths on CREW PRAM which runs in $O(n)$ time with n intervals in a graph. We have given a linear processor CREW PRAM algorithm for determining the shortest paths in an interval graphs.

Chapter 4 - A new method for detecting low dimensional chaos in noisy small sample sets is presented. A quantity that can be interpreted as a measure of the degree of determinism or nonlinear mean predictability in a time series is defined on the basis of the embedding theorem and the method of time delays (Takens 1981). Numerical experiments on stochastic and chaotic processes show that, in application to very short time series, this method is effective, while traditional approaches such as the false nearest neighbors method have difficulties.

Chapter 5 - This short report deals the use of three dimensional CFD (Computational Fluid Dynamics) simulations in order to better understand the complex interactions between the hydrodynamics of a given water body and the chemical, physical, biological and meteorological phenomena.

Chapter 6 - Freud's interpretation of the Oedipus story was the cornerstone of classical psychoanalysis, leading early psychoanalysts to seek repressed wishes among patients to kill their fathers and mate with their mothers. This literal interpretation overlooks a key feature of the Oedipus story, the riddle of the Sphinx. This paper re-examines the Sphinx's riddle – "What walks on four legs in the morning, two legs at noon, and three legs in the evening?" – as a paradox of self-reference. The riddle is paradoxical, by seeming to contradict all known laws of science, and self-referential, in that its solution depends upon Oedipus applying the question to himself as a human being. By threat of death, Oedipus must understand that morning, midday and evening refer not literally to one day, but metaphorically to stages of life. This paper links ancient myth with contemporary computational studies by interpreting the capacity for self-reference as a Universal Turing Machine with full memory, both implicit and explicit, of its own past. A cybernetic perspective dovetails with research on the neurobiology of memory, as well as with cognitive studies derived from developmental psychology. Mental skills required for self-reference and metaphorical thinking signal internal complexity and mature cognition necessary to enter the arena of modern self-reflective consciousness.

Chapter 7 - In this letter, we address the time scale synchronization between two different chaotic systems from the view point of empirical mode decomposition and the results are compared with those obtained using wavelet theory. In empirical mode decomposition method, we decompose a time series in distinct oscillation modes which may display a time varying spectrum. In this process it was observed that the transition of non synchronized, phase synchronization, lag synchronization and complete synchronization occurs for different coupling parameter values. The quantitative measure of synchronization for empirical mode decomposition and wavelet approach has been proposed. It has been observed that due to the presence of a scaling factor in the wavelet approach it has more flexibility for application.

Chapter 8 - We develop a non linear methodology to analyze nuclear medicine images. It holds on Recurrence Quantification Analysis (RQA), on Analysis of Variability by variogram, and on estimation of Fractal Dimension. It is applied to five subjects with osteoporosis in comparison with five control subjects. Bone nuclear images are obtained after administration of ^{99m}Tc-HDP. Regions of interest (ROI) are selected in maxillary bones of oral cavity. Some basic non linear indices are obtained as result of the methodology and they enable quantitative estimations at the micro-structural and micro-architectural level of bone matrix investigated. The indices result very satisfactory in discriminating controls from subjects with osteoporosis. Their appear of interest also in the case of dentist intervention often engaged to evaluate oral signs and, in particular, to utilize mandibular or maxillary bones indices in relation to possible loss of bone mineral density and/or to microarchitectural deterioration of bone tissue.

Chapter 9 - In the last years, growing attention has been paid to the reconstruction of chaotic attractors from one or more observables. In this paper a Multi Layer Perceptron with a tapped line as input, is used to forecast the hypercaotic Rössler system state variables starting from measurements of one observable. Results show satisfactory prediction performance if a sufficient number of taps is used. Moreover, a sensitivity analysis has been performed to evaluate the predictiveness of the different delayed input in the neural network model.

Chapter 10 - Fractal geometry is also known as "Mandelbrot's geometry" in honour to its "father" the mathematician Benoit Mandelbrot (b. 1924), that showed how fractals can occur in many different places in mathematics and in other disciplines. Fractal geometry can be used for modelling natural shapes (e.g., fern, trees, seashells, rivers, mountains), and its important applications also appear in computer science, because this "new" geometry permits to compress the images, to reproduce, in the virtual reality environments, the complex patterns and the irregular forms present in nature using simple iterative instructions. The aim of this paper is to present the applications of fractal geometry in computer graphics (to compress images) and in virtual reality (to generate virtual territories and landscapes).

Chapter 11 - This article: 1) develops new psychological theories and mathematical models that can explain many of the legal and economic problems that occurred in the US housing industry between 2000 and the present – such as the sub-prime loan problems, predatory lending, mortgage fraud, title defects, rapid/un-warranted price increases and sales fraud; 2) analyzes and identifies the psychological and behavioral biases of first-time homebuyers and repeat home buyers, 3) develop new theories (testable hypothesis) of psychological effects and Biases inherent in the housing purchase/sale process; 4) develops theoretical mathematical models for Buyers' Propensity-To-Purchase.

This study involves analysis of historical economic trends, critique of existing methods and theories, and development of new theories and development of mathematical models. This article also partly relies on surveys and published empirical research using US macroeconomic and housing data from the 1995-2003 period. At the present time, the models developed in this article cannot be realistically tested empirically because real estate data and price series and psychological effects described in the models (and associated periodic changes in such data or the logarithms of such data) don't fit known distributions and regression techniques.

Chapter 12 - The climatic memory of 5 deep lakes (Maggiore, Como, Garda, Iseo and Orta) shows a series of warming and cooling phases from 1887 to 2007 that cannot be attributed to the energy exchanges at the air-water interface alone. This underlines the complexity of the lakes ecosystems' response to the ongoing global change.

Chapter 13 - The Economist magazine of May 23rd 2002, was featuring a special section: "People already worry about genetics. They should worry about brain science too". The cover was about the fear of a near-future of *mind control*: "If asked to guess which group of scientists is most likely to be responsible, one day, for overturning *the essential nature of humanity*, most people might suggest geneticists. In fact neurotechnology poses a greater threat, and also a more immediate one. Moreover, it is a challenge that is largely ignored by regulators and the public, who seem unduly obsessed by gruesome fantasies of genetic dystopias." The journalistic emphasis might be criticized from many points of view, for example: who knows what is the *essential nature of humanity*? Anyway, as mind sciences are progressing, there are several new issues on free will and personal responsibility which are worth some reflections.

Progress in Chaos and Complexity Research
Editors: Franco F. Orsucci and Nicoletta Sala
ISBN: 978-1-60456-375-7

Chapter 1

DETECTION OF TRANSIENT SYNCHRONIZATION IN MULTIVARIATE BRAIN SIGNALS APPLICATION TO EVENT-RELATED POTENTIALS

***Axel Hutt*[1]* *and Michael Schrauf*[2]†
[1]Institute of Physics, Humboldt University of Berlin
Newtonstr. 15, 12489 Berlin, Germany
[2]DaimlerChrysler AG, Research and Technology
Information and Communication, 096 / T 728 - RIC/AP
Hedelfinger Strasse 10-14, 73734 Esslingen, Germany

Abstract

The present work introduces an analysis framework for the detection of transient synchronized states in multivariate time series. In case of linear data these states exhibit a dramatic increase and subsequent decrease of time scale in the signal, whereas the corresponding instantaneous phases show transient mutual phase synchronization. We propose a single segmentation algorithm for both data types, which considers the space-time structure of the data. Applications to linear and phasic simulated signals illustrate the method. Further applications to event-related brain potentials obtained from an auditory oddball experiment while real car-driving reveal the lack of the cognitive component P300 in an experimental condition. Further the obtained results indicate attention effects in the event-related component N100 and shows dramatic latency jitters between single datasets. A comparison of the proposed method to a conventional index of mutual phase synchronization demonstrates the superiority of considering space-time data structures.

1. Introduction

In the last decades synchronization has been found in various systems in biology, physics or medicine (Pikovsky et al. (2001)). In neuroscience, synchronization has attracted much attention as a concept of information processing in the brain (see e.g. Singer & Gray (1995)).

*E-mail address: Axel.Hutt@physik.hu-berlin.de
†E-mail address: michael.schrauf@daimlerchrysler.com

This approach is in the tradition of results found more than two decades earlier, which revealed cooperative, i.e. synchronized, activity in spatial cortical columns (Wilson & Cowan (1972); Luecke & von der Malsburg (2004)). In addition, several studies have shown strong correlations between cooperative dendritic activity of neurons and electromagnetic activity on larger spatial scales, e.g. local field detectors or encephalographic potentials and fields (Freeman (2000); Nunez (1995)). The present study focus to synchronization effects of these evoked electroencephalographic potentials.

There are several crucial aspects in the analysis of multivariate evoked potentials, which have to be considered. The most important ones are explained in the following. In neuropsychology, most experiments apply paradigms with several different conditions to gain information about a specific functionality of the brain, as e.g. processing of semantic differences (Kotz et al. (2001)) or prosody in stimuli (Schirmer et al. (2002)). In order to gain significant results, single experimental conditions are repeated several times. The number of these repetitions depends on the complexity of the task and is typically in the range of $50 - 500$. In case of rather complex experimental paradigms, the number of trials is low, i.e. in the range of $10 - 50$. To extract significant results, trials of the same experimental condition are averaged. This procedure is reasonable under the assumption of seldom artifacts, as head movements or low attentiveness of the subject. However, in contrast to most experiments under controlled conditions in a labortaory, more and more experiments are carried out under less controlled every-day-life conditions (Schrauf & Kincses (2003)). In these cases, only few repeated trials are acquired and artifacts play an important role. To extract significant results anyway, the analysis of averages over few trials or even of single trials is necessary. Several corresponding methods have been proposed in the last years (Laskaris & Ioannides (2002); Ioannides et al. (2002); Karjalainen & Kaipio (1999); Lachaux et al. (2002)). In the subsequent sections, we aim to develop a method to gain information on synchronization in single trials.

Further, one of the major aims of multivariate analysis in neuropsychological research is the detection of functional components from observed data. Lehmann & Skrandies (1980) developed an algorithm to extract spatial activity maps from single data sets, which show metastable synchronized behaviour in time. The major idea of the algorithm consists in comparing the spatial distributions of brain activity on the scalp of successive time points and thus is succesfull mainly for averaged data sets, which exhibit smooth behaviour. These obtained time segments of similar activity distributions are called microstates and reflect functional states in the brain (Brandeis et al. (1995)). Subsequent work of Pascual-Marqui et al. (1995) and Wackermann (1999) extended this approach by a cluster algorithm and a classification scheme of the extracted components, respectively. These approaches are more robust towards noise in the data. However, the method applies the cross-validation method to determine the number of clusters and fails for high dimensional data. The proposed method extend the latter approaches by a clustering algorithm, which also extracts the time segments from the data but solves the problem of the number of clusters additionally. This will be shown in the subsequent sections.

Previous studies have attacked the latter problems, i.e. multiple trials and optimal choice of the time window, separately in the context of synchronization. However, as far as we know, both problems have not been attacked in a common approach. In this context, both linear and phasic multivariate brain signals attracted attention. The former represents the observed data itsself while the latter represent the instantaneous phases extracted from the linear data. The two best-known definitions of instantaneous phases are given by the Hilbert- and the wavelet transformation. Recently, the analysis of phase synchronization between single time series attracted increased attention (Tass (1999); Haig et al. (2000); Lee et al. (2003); Lachaux et al. (1999)). However, applications to typical encephalographic data need to consider a large set of spatially-distributed detectors as microscopic generators spread their activity on the scalp. Some methods have been developed to extract instantaneous mutual phase synchronization (Haig et al. (2000); Rosenblum et al. (2000)). However, these methods neglect the spatial distributions of phases. In addition, we mention the work of Allefeld & Kurths (2003); Allefeld et al. (2005), who recently developed a method which extracts an instantaneous index for global phase synchronization by considering the space-time structure of data. However, this method as most of the previously developed methods considers a high number of trials and thus is not able to examine single data sets. That is to our best knowledge the detection of mutual phase synchronization in single data sets is still lacking.

The present work proposes a novel synchronization analysis for both linear and phasic data sets. It extends both algorithms detecting global phase synchronization in multiple trials to single trial analysis and the analysis of linear single trial analysis to the treament of phasic data. Our approach considers the spatiotemporal behaviour of multivariate brain signals and aims to extract time segments of transient synchronization. The key point of the proposed approach is the observation that, in some time segments, all time series in a dataset show a mutual change of their time scale. This behaviour is well-known from studies of encephalographic brain activity (Lehmann & Skrandies (1980); Freeman (2003); Kay (2003); Tsuda (2001); Uhl et al. (1998); Breakspear & Friston (2001); Hutt & Riedel (2003)) and yields clusters in the high-dimensional data space. This observation is valid for both linear (Hutt & Kruggel (2001)) and phasic data (Hutt et al. (2003)), while clusters in linear data represent mutual synchronization and clusters in phasic data represent mutual phase synchronization. Hence, a single algorithm, which extracts clusters in data space, is capable to detect synchronized mutual activity. The distinction of linear and phasic data comes in by the data topology, i.e. linear data behaves on a plane and phasic data lives on a torus.

The present work is structured as follows. The next section 2. introduce the cluster algorithm for both linear and phasic data, discusses the applied statistical evaluation and introduce the examined data. The latter represents auditory evoked potentials obtained experimentally during a real car-driving experiment. Applications to simulated linear and phasic signals in section 3. illustrate briefly the method, while the extensive application to the experimental data reveal the effects of averaging of trials and the latency jitter of temporal segments in single trials. Further, typical functional components are detected, while the component N100 exhibits latency differences in two different experimental conditions. This finding indicates early cognitive processing in the brain after 100ms from stimulus onset. The subsequent discussion in section 4. closes the work.

2. Methods

2.1. Clustering of Linear Data

Let us consider two typical time series $Q_{Fz}(t)$, $Q_{Cz}(t)$ obtained during a cognitive experiment (Fig.1(a)). We observe mutual behaviour of the time series about 105ms, 276ms and 331ms, i.e. the time series exhibit a minimum or maximum. These extrema in single time series are interpreted in neuropsychology as indicators to neural functional processes (Rugg & Coles (1996)).

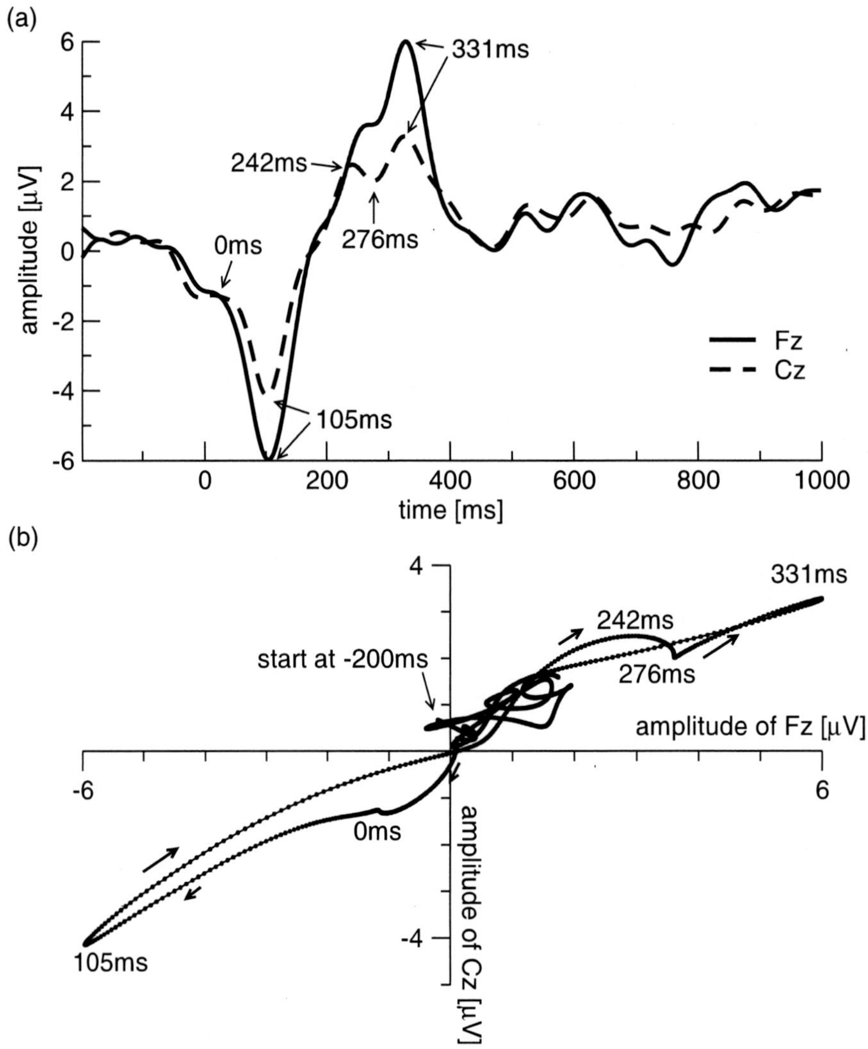

Figure 1. Two typical time series of observed electroencephalographic potentials, here taken at detectors Fz and Cz. They are plotted as single time series (top part) and trajectory in data space (bottom part). The arrows in the bottom part denote the temporal evolution direction of the signal.

Now applying an approach from physics, Figure 1(b) shows both time series as a trajectory in data space. Obviously, the trajectory at the three time points exhibits turning points. Focussing to these turning points, the data around the turning points are more dense than between these points. We may say the time scale of the$N$-dimensional signal decreases near turning points. Since turning points of trajectories exhibit vanishing temporal derivatives additionally, it is for M time series

$$\frac{dQ_k(t)}{dt} = 0 \quad \forall k = 1,.....,M.$$

with $t \in I$. Here, I denotes the time interval near turning points. For a N-dimensional signal, $M = N$ would mean a mutual synchronization of all time series, while in the case of $M < N$ only M time series are synchronized. We call this synchronization effect mutual signal synchronization (MSS), which is similar to the mutual synchronization introduced in previous works (Pikovsky et al. (2001); Stam & Dijk (2002); Breakspear & Terry (2002)). Further, the subsequent sections shall introduce an quantity for mutual synchronization of signals instantaneous in time, whose value gives the degree of MSS. That is, low values indicate mutual behaviour with small$M$, whereas high values indicate high values of $M \approx N$. We point out, that we hesitate to use one of the well-known synchronization definitions (Pikovsky et al. (2001)), as we can not say anything about the underlying generating mechanisms.

Indeed, the phenomenon of inreased data density has been found in multivariate data implicitely in various studies (Lehmann & Skrandies (1980); Pascual-Marqui et al. (1995); Hutt (2004); Hutt et al. (2000); Hutt & Kruggel (2001); Hutt & Riedel (2003)). In case of event-related potentials/event-related fields (ERP/ERF), such metastable phenomena have been called differently in literature, e.g. *microstates* by Lehmann & Skrandies (1980), *quasi-stationary states* (Hutt & Riedel (2003)), states of *event-related synchronization* (Pfurtscheller & da Silva (1999)) or *event-related components* in many neuropsychological studies (see e.g. Rugg & Coles (1996)). In addition, we mention the notion of chaotic itinerancy (Tsuda (2001); Kay (2003)), which models the transients by phase transitions of first order (see e.q. Freeman (2003)). Although there are differences in these approaches, they describe the mutual decrease and subsequent increase in the time scale of data. In addition, all definitions classify such metastabilities by their latency shift from stimulus onset and their spatial activity distribution at the corresponding latency. In the following, we shall call these mutual phenomena simply components. Hence re-considering the previous discussion components reflect mutual synchronization.

In case of non-smooth data, the occurrence of mutual behaviour of time series is not that obvious anymore, but trajectory segments of components still exhibit an increased data density. In mathematical terms, turning points subject to noise obey

$$Q_i(t) = \bar{Q}_i + \Gamma_i(t) \quad \forall i = 1,...,N$$

where the deterministic part $\bar{Q}_i$ evolves on a much large time scale than the random fluctuations Γ_i, which exhibit $\langle \Gamma_i \rangle = 0$. Here $\langle ... \rangle$ denotes the ensemble average. Hence, at a constant sampling rate trajectories near turning points obey$\langle Q_i(t) \rangle \approx \bar{Q}_i$ and, subsequently, components represent agglomerations in of data points in data space, i.e. clusters.

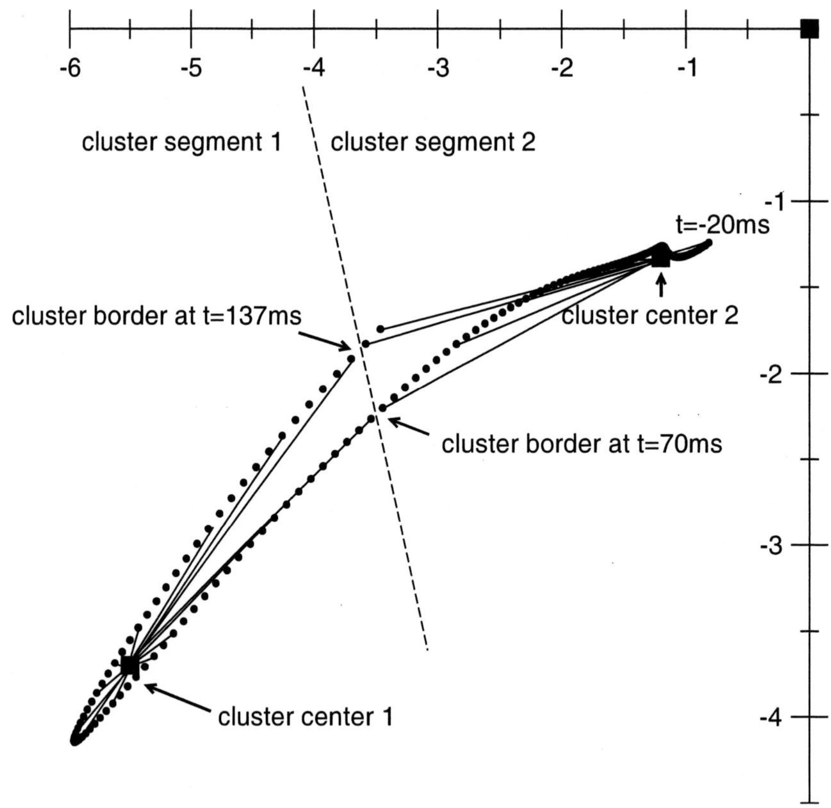

Figure 2. A trajectory segment in the time window $[-20\text{ms}; 139\text{ms}]$. The dashed line illustrates the border of both clusters.

To detect these clusters, we apply the K-means cluster algorithm (Duda & Hart (1973)) which assumes a priorily a fixed number of clusters. Figure 2 shows a trajectory segment extracted from the data in Fig. 1. Two cluster centers have been guessed for illustration reasons. Here data between -20ms and 70ms and the two last data points are nearer to cluster center 2 than to cluster center 1, while the data between 71ms and 137ms belongs to cluster 1. That is the two cluster centers segment the data into three temporal intervals, whose borders at 70ms and 138ms are determined by the distance from cluster centers. Now, we apply the K-means algorithm to the data segment of Fig. 2 for $K = 2, K = 3$ and $K = 5$ clusters, respectively. Figure 3 shows the computed squared Euclidean distances from cluster centers to data for the different number of clusters and the plots exhibit the change of nearest clusters and subsequently temporal segments.

The proposed method aims to find a reasonable quantity that distinguishes well-separated from intersecting clusters while taking into account errors by single outliers. This quantity represents the cluster quality of a data point at time t and is defined by the area $a_l(t)$ in Fig. 3, while $l = 1..N_K$ and N_K denotes the number of segments for a fixed number of clusters K. This area between the nearest and the second nearest cluster quantifies both the spatial separation of two segments and its cardinality.

In mathematical terms, the well-known global cost function for K-means clustering and

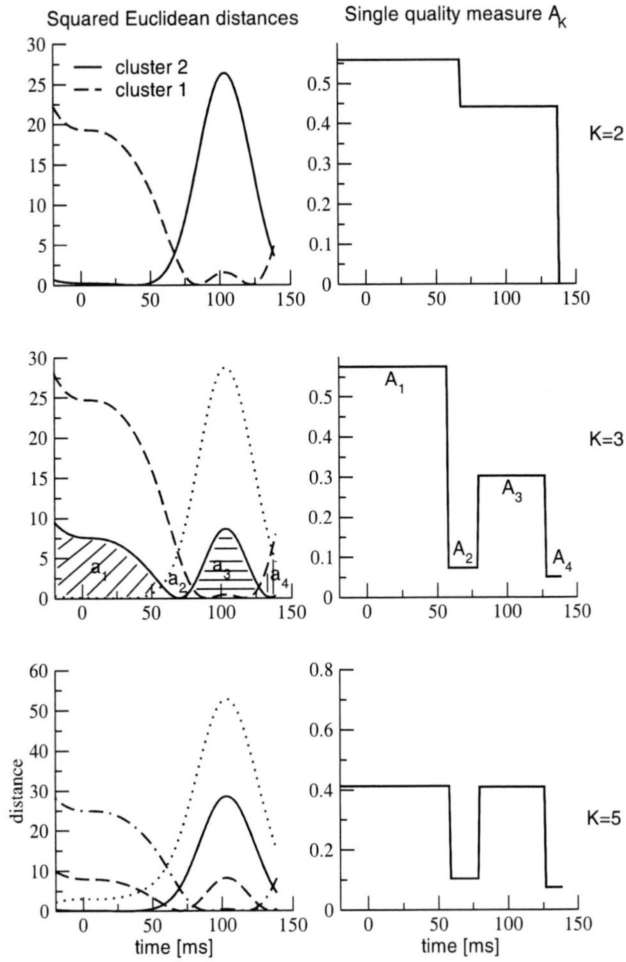

Figure 3. The basic elements of the introduced cluster quality illustrated for number of cluster $K = 2,\ K = 3$ and $K = 5$.

K clusters reads

$$V_K = \sum_{l=1}^{K} \sum_{i \in C_l} (\mathbf{x}_i - \bar{\mathbf{x}}_l)^2 = \sum_{l=1}^{K} \sum_{i \in C_l} d_{ik}^2 \tag{1}$$

where $\bar{\mathbf{x}}_l$ denote cluster centers and C_l are the corresponding sets of members. V_K gives the mean distance of data to clusters and is minimal for the optimal choice of cluster centers. According to the previous discussion, the method extends this formulation to temporal segments S and also considers the distance to the second nearest cluster of each data point. That is

$$V_K' = \sum_{l=1}^{S} \sum_{i \in S_l} (e_{il}^2 - d_{il}^2) = \sum_{l=1}^{S} (N_l - 1)(\sigma_l^{sn} - \sigma_l^n) = \sum_{l=1}^{S} a_l \tag{2}$$

where d_{ik} and e_{ik} denote the Euclidean distance from the data point $\mathbf{x}_i$ to its corresponding nearest and second-nearest cluster center in segment l, respectively. N_l represents the number of data in segment l. Here, a_l is proportional to the difference of cluster variances σ_l^{sn} and σ_l^n between second-nearest and nearest cluster center in segment l,respectively. Now, in contrast to the global approach in (1),(2), the method associates each data point i to the cluster quality of its segment by $A_{li}^{(K)} = a_l I[i]$ with the indicator function $I[i \in S_l] = 1$, $I[i \notin S_l] = 0$. Since the K-means algorithm assumes a fixed number of clusters to be detected, the superscript (K) makes clear that $A_{li}^{(K)}$ represents the cluster quality for a fixed number of clusters K. Finally, the normalization of $A_{li}^{(K)}$ and averaging over increasing number of clusters, i.e.

$$\bar{A}_l^{(K)}(i) = \frac{A_{li}^{(K)}}{\sum_{l=1}^{S} A_{li}^{(K)}} \quad , \quad p(i) = \frac{1}{U-2} \sum_{K=2}^{U} \bar{A}_l^{(K)}(i)$$

yields the mutual signal synchronization index $p(i)$ for each sample point i. Here, U is the maximum number of clusters set to $U = 20$ in the present work. Previous studies (Hutt & Riedel (2003)) showed that results are robust towards the value of U if U exceeds the maximum number of expected clusters. According to this definition, large values of p give well-separated clusters, i.e. well-detected components, while falls and rises mark transitions between different clusters. In the following, the synchronization index $p(t)$ for linear data is called mutual synchronization index $MSS(t)$.

2.2. Clustering of Circular Data

In addition to the analysis of linear data, this section treats phasic or circular data. Several previous studies examined phase synchronization in evoked brain signals (Tass (1999); Allefeld & Kurths (2003); Allefeld et al. (2005); Haig et al. (2000); Breakspear (2002)). Since a previous theoretical study has shown increased data densities in temporal segments of mutually phase-synchronized data (Hutt et al. (2003)), the extention of the derived cluster algorithm to circular data is straight forward.

Phasic data make sense in a physics point of view only in an associated narrow frequency band, as phases are strongly related to their temporal change, i.e. the frequency (Pikovsky et al. (2001)). Now to obtain instantaneous phases from linear data, the present work applies a Gaussian filter in frequency space in combination with a complex Fourier transform (DeShazer et al. (2001)) obtaining

$$S(t) = 2\int_{-\infty}^{\infty} e^{-(\nu-\nu_k)^2/\sigma_\nu^2} \tilde{Q}(\nu) e^{-i\nu t} d\nu. \tag{3}$$

Here, $\tilde{Q}(\nu)$ denotes the Fourier transform of the signal $Q(t)$. Since the center frequency holds $\nu_k > 0$, $S(t)$ is complex and the instantaneous spectral power and phase is given by

$$A(t) = \sqrt{\mathcal{I}(s(t))^2 + \mathcal{R}(s(t))^2} \quad , \quad \Phi(t) = \arctan \frac{\mathcal{I}(s(t))}{\mathcal{R}(s(t))} \tag{4}$$

for each frequency band about ν_k, respectively. Here, $s(t) = S(t) - \bar{S}$, $\bar{S}$ is the temporal average of $S(t)$ and $\mathcal{R}(s)$ and $\mathcal{I}(s)$ denote the real and imaginary part of s, respectively. The width of the frequency band is given by the variance of the filter σ_ν^2, which in turn determines the variance of the corresponding temporal filter by $\sigma_t^2 = 1/\sigma_\nu^2$ according to the uncertainty principle. The corresponding standard deviation in the time domain represents an estimate for the number of correlated time points and we fix it to 10 oscillations, i.e. $\sigma_t = 10/\nu$. Subsequently, filtered data in low frequency bands exhibit higher temporal correlations than data for higher frequencies. In turn, the width of the frequency filter is proportional to the center frequency by $\sigma_\nu = \nu/10$. We mention the equivalence of this approach to the analysis by Morlet wavelets.

According to Pikovsky et al. (2000), mutual phase synchronization (MPS) exhibits bounded differences of phase pairs

$$|(\Phi_k(t) - \Phi_l(t)) \bmod 2\pi| < \text{const} \quad \forall \quad k = 1, ..., N, \; l = k, ..., N.$$

Hence MPS yields data clusters in the extended space of all phase pairs defined by a new multivariate time series $\mathbf{y}(t) \in \mathcal{R}^M$ with $M = N(N-1)/2$ and $\{y_j(t)\} = \{\Phi_k(t) - \Phi_l(t) \; \forall \, k > l\}$.

There are just two more implementation differences to the linear case, namely the computation of circular distances and the computation of mean circular values. These computations obey basic rules in circular statistics and we refer the reader to Mardia & Jupp (1999) for more details. All subsequent computations of distances, averages and variances of circular data obey these rules.

Summarizing the proposed method for circular data, at first choose a narrow frequency band, then compute the circular time series by (3) and (4) and compute the new extended data set of phase differences before applying the cluster algorithm as proposed in the previous section.

Similar to the linear case, the obtained cluster quality exhibits large values in case of strong MPS, while sharp falls and rises, respectively, mark transitions between different clusters. In the following, the synchronization index $p(t)$ for circular data is called mutual phase synchronization index $MPS(t)$.

2.3. Statistical Analysis

Since the K-means algorithm is iterative and the obtained cluster centers are sensitive to initial values, there is no guarantee that the algorithm converges to the optimal cluster results. Hence, we repeat the computation of $p(t)$ 10 times obtaining mean values $P(t)$ and corresponding variances $\sigma(t)$ for each time point t. To assess additionally the cluster results, surrogate data are generated by randomizing the data in time and the re-application of the cluster algorithm yields new mean cluster qualities $P_s(t)$ and corresponding variances $\sigma_s(t)$. The obtained surrogate data set exhibits a decorrelated temporal structure. Subsequently, no prominent cluster segment occurs and $P_s(t)$ is much smaller than in the original data. We shall verify the missing temporal struture by visual inspection, while the lower values of P_s are verified by the t-test for every time point t. The t-value reads

$$T(t) = \frac{P(t) - P_s(t)}{\sigma(t) + \sigma_s(t)} \sqrt{n}, \tag{5}$$

with the degrees of freedom $n = 19$. Equation (5) sets the null hypothesis such that P is indistinguishable from random cluster results P_s. For $T(t) > t_{\alpha,n}$ the test rejects the null hypothesis at an false positive error rate α, that is P is significantly different from P_s. Here $t_{\alpha,n}$ denotes the Students t-distribution. In the following, we set $\alpha = 0.05$.

In addition, the present work considers a mutual phase synchronization index motivated by Haig et al. (2000) and applied recently by Allefeld & Kurths (2003). This index is the global circular variance (Mardia & Jupp (1999))

$$R(t) = \frac{1}{L}\sum_{l=1}^{L}\sqrt{\left(\sum_{j=1}^{M}\sin y_{jl}(t)\right)^2 + \left(\sum_{j=1}^{M}\cos y_{jl}(t)\right)^2}\,, \tag{6}$$

where $\{y_{jl}\}$ are phase differences in trial $l = 1\ldots L$. $R(t)$ gives a rough estimate of MPS for each frequency band. This index extracts information from trial ensembles and is not applicable for single trial analysis. However, we shall compare our results on single trial averages to results from Eq. (6) in a later section.

2.4. Data Acquisition

Event-related potential (ERP) data are analyzed in two conditions of a 2-tone passive oddball paradigm. Tones used were a standard at a frequency of 1kHz and an occurence rate of 0.85 and a deviant tone at a frequency of 2Khz and an occurence rate of 0.15. The tones were presented at a level of 70 dBSL and had a rise and fall time of 10ms, a duration of 50ms with an inter-stimulus-interval from 3.2s to 3.8s between the start of each stimulus. Tones were played through earphones. ERP recordings were made from 32 sites (electro-cap, 10 : 20-system, impedance $< 5k\Omega$, linked mastoid reference) at a sampling rate of 1kHz and amplitude resolution $0.1\mu V$. Hardware filters were applied with the low cutoff at 0.5Hz, the high cutoff at 70Hz and the notch filter at 50Hz.

Topographical scalp current source density(CSD)-maps (order of splines:4, max. degree of Legendre polynomials: 10) were made for comparisons. The frontal (Fz), central (Cz), and parietal (Pz) midline electrode sites were used to facilitate correct identification of the P300 peak (Johnson (1993)). EOG artifact rejection was applied (Gratton et al. (1989)). Data were evaluated offline using a digital low-pass 25Hz filter (e.g. Polich (1998)). Driving tasks (with or without using an active cruse control named distronic) were alternated every 30 min to minimize effects of sequence and attention.

Recordings were analyzed from one physically and mentally healthy subject (male, 45 years, 25 years driving experience, about 50.000 km driven with the Mercedes Benz S-500 test car), with no history of neurological disorder, free of medication and corrected to normal vision. The test route was a 400 km stretch of a german highway (Stuttgart-Duesseldorf). Digital video of forward road scene was recorded for comparison of traffic density and to identify particular variations of traffic scenes.

3. Results

Now, we examine synchronization results both from simulated and empirical multivariate signals. In the latter examination, both linear and circular data shall be examined for all experimental conditions. Since the present work proposes an algorithm to examine single data sets, we show results from averages over all trials and from averages over subsets of trials.

3.1. Application to Simulated Linear Data

This section illustrates features of the introduced mutual synchronization index MSS by application to artificial multivariate data. The investigated dataset $\mathbf{q}(t)$ represents a superposition of three interacting modes

$$\mathbf{q}(t) = x(t)\mathbf{v}_x + y(t)\mathbf{v}_y + z(t)\mathbf{v}_z,$$

where amplitudes $x(t), y(t), z(t)$ obey a 3-dimensional dynamical system

$$\begin{aligned}\frac{dx}{dt} &= x - x(x^2 + 4y^2) + \Gamma(t)\\ \frac{dy}{dt} &= y - y(y^2 + 4z^2) + \Gamma(t)\\ \frac{dy}{dt} &= z - z(z^2 + 4x^2) + \Gamma(t)\end{aligned} \tag{7}$$

with identically-distributed noise $\Gamma(t) \in [-0.1; 0.1]$. Dynamics described by Eqs. (7) arises in various physical systems, e.g. in rotating fluids at large Taylor numbers (Busse & Heikes (1980)).

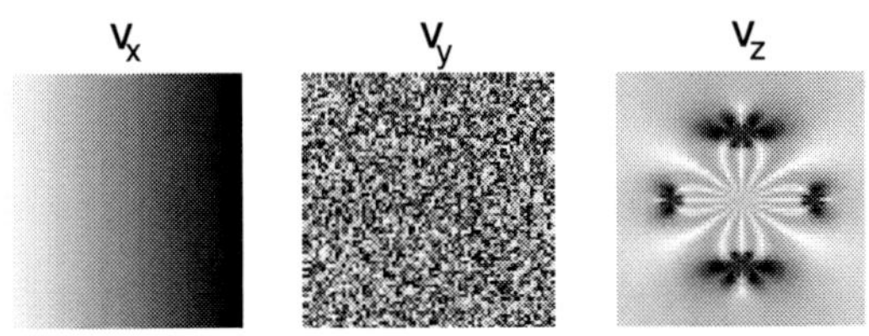

Figure 4. Spatial modes of the simulated data.

In the current context, spatial modes $\mathbf{v}_x, \mathbf{v}_y, \mathbf{v}_z$ represent artificial 75x75-patterns (Fig. 4), i.e., the signal $\mathbf{q}(t)$ lives in a 5625-dimensional space. It is generated by 2200 integration steps with initial conditions $(x(0), y(0), z(0))^t = (0.03, 0.2, 0.8)^t$ and its trajectory passes saddle points at $\mathbf{x}_3 = (0,0,1)^t, \mathbf{x}_1 = (1,0,0)^t$, $\mathbf{x}_2 = (0,1,0)^t$ and $\mathbf{x}_3 = (0,0,1)^t$ in this sequence. Fig. 5 shows a sampled time series of $\mathbf{q}(t)$.

In Fig. 6, the mutual synchronization index MSS shows plateaus, troughs and steep rises. Large values of MSS originate from large values of $\overline{A_l^{(K)}}(i)$ (cf. section 2.1.) and indicate clusters in data space, whereas rapid changes reflect points with changing cluster memberships for different K. Thus, plateaus represent synchronized states, while troughs and rises mark their upper and lower borders, respectively.

Since spatio-temporal clusters are specified in temporal and spatial domains, averages of data in synchronized states yield cluster centers in data space, i.e. 75x75-patterns. Fig. 7 shows the computed cluster centers in the time intervals $[0; 314]$ (cluster 1), $[416; 929]$ (cluster 2), $[1150; 1450]$ (cluster 3) and $[1710; 2130]$ (cluster 4). These patterns show good accordance to original patterns (Fig. 4).

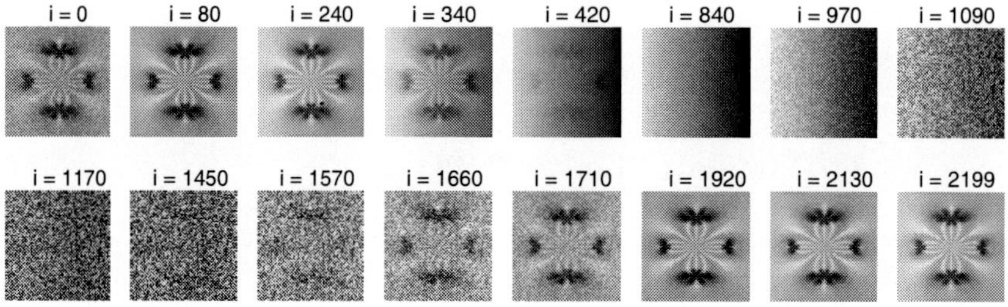

Figure 5. Sampled time series of spatial maps of the simulated data. Quasi-stationary states emerge at $i \approx 240$, 840, 1450 and $i \approx\sim 1920$.

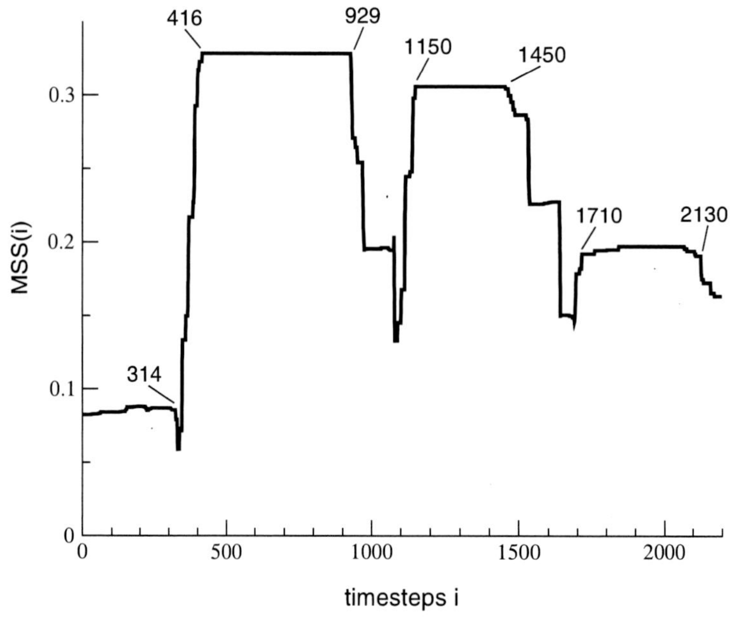

Figure 6. Clustering results of simulated data in the time window $[0;2199]$. The mutual signal synchronization index $MSS = p$ is plotted with respect to timepoints i. Plateaus, i.e. metastable states, occur at $[0;314], [416;929], [1150;1450]$ and $[1710;2130]$.

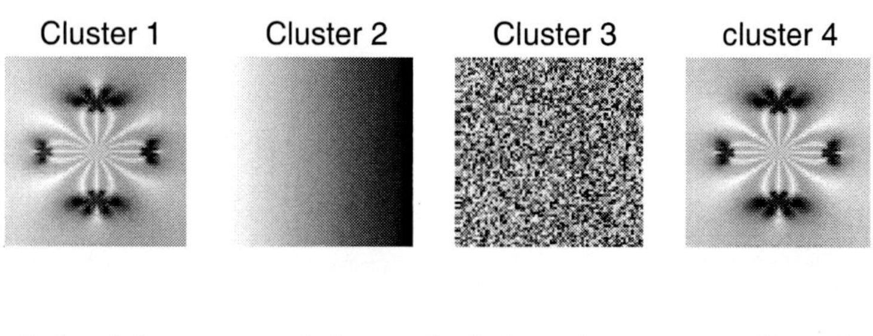

Figure 7. Spatial averages of clustered windows in corresponding sequence.

3.2. Application to Simulated Circular Data

Evaluating our method by application to multivariate circular data, now we study phase signals obtained from chaotic data. Note that phase synchronization always occurs with respect to a phase reference. In general, however, multivariate signals do not serve a unique reference and, therefore, we examine data sets including all couples of phase differences. The system in question is a ring of 5 uncoupled Lorenz systems

$$\dot{x}_i = -10x_i + 10y_i \quad (8a)$$

$$\dot{y}_i = 28x_i - y_i - x_i z_i + C(y_{i+1} + y_{i-1} - 2y_i) \quad (8b)$$

$$\dot{z}_i = x_i y_i - \frac{8}{3} z_i + F(t) \quad , \quad i = 1, \ldots, 5 \quad (8c)$$

driven by an external force $F(t) = 10 \cdot \sin(8.3t)$. This system yields so-called imperfect phase synchronization (Zaks et al. (1999)), i.e. the phase of every single attractor drifts in short segments by multiples of 2π relative to the external force $F(t)$. This drifting is caused by the broad range of intrinsic time scales of the Lorenz system. We realized numerical solutions of Eqs.(8) by applying an Euler-forward algorithm with step size 0.01, where uniformly distributed initial values $(x_i(0), y_i(0), z_i(0)) = (8.4 + \Gamma_x, 8.4 + \Gamma_y, 40 + \Gamma_z)$ with $\Gamma_{x,y,z} \in [-0.5; 0.5]$ guaranteed a stable integration of $T = 1500$ time steps.

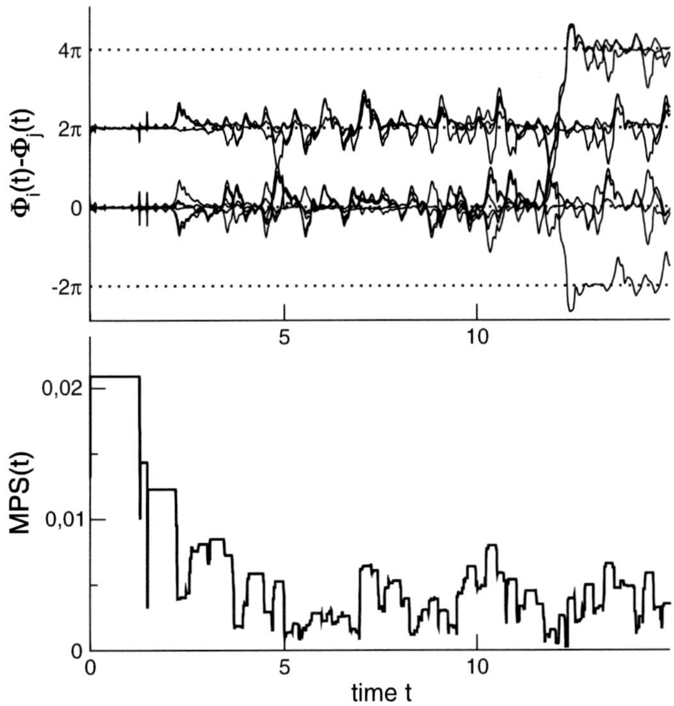

Figure 8. Time series of phase differences and the computed synchronization index for un-coupled Lorenz systems driven by an external force.

Figure 8, top panel, displays time series of 10 phase pairs $\Delta\Phi_{ij} = \Phi_i - \Phi_j$ obtained from the of amplitudes $\{y_i\}$. Only a short plateau of the synchronization index occurs (Figure 8, bottom panel). This behaviour originates from the similar initial values of the 5 Lorenz systems, which thus synchronize briefly at the beginning of the simulation. After this initial transient part, the Lorenz systems desynchronize and $MPS(t)$ drops to low values. Here, it is $p < 0.001$, i.e. the results are highly significant.

3.3. Application to Experimental Linear Data

Now, the experimental brain data are examined. Figure 9 presents time series from averages over all trials for both experimental conditions. Conventional methods, i.e. peak detection in single time series, identify the components N100 at $\sim$ 100ms, component P200 at $\sim$ 200ms and component P300 at $\sim$ 300ms for the distronic condition. For the non-distronic condition, we identify the components N100 at $\sim$ 100ms and P200 at $\sim$ 200ms, while no component P300 is observed.

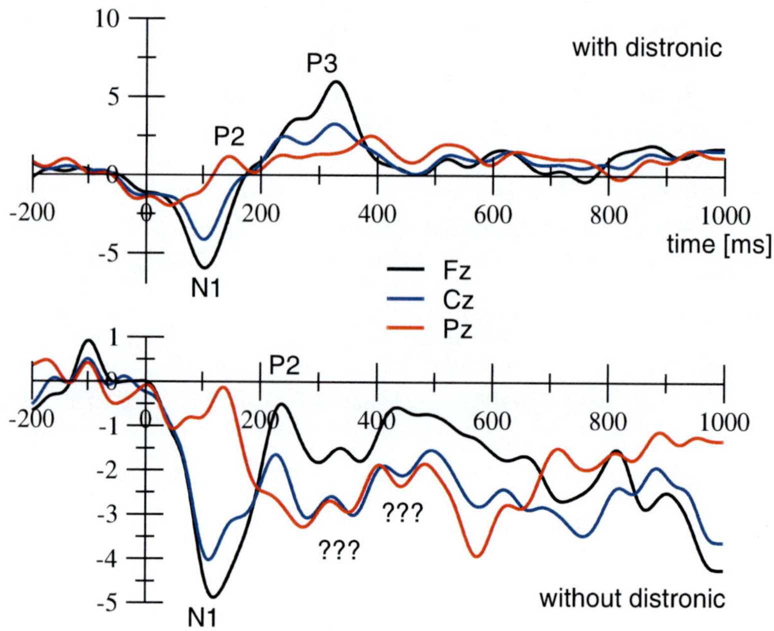

Figure 9. Observed time series at different detectors for both experimental conditions. Conventional methods classify the components N100, P200 and P300 for the distronic condition and N100 and P200 for the non-distronic condition at corresponding temporal latencies.

The proposed method yields the mutual synchronization index $MSS(t)$. First, the mutual synchronization index is extracted from each single trial. A subsequent average over all obtained results yields plateaus of constant values with sharp edges. Figure 10a shows the averages over all trials for both experimental conditions. For the time window $[-200ms; 1000ms]$, i.e. the whole signal, we observe various time segments with large values of MSS, which coincide to the conventional results from Fig. 9. Focussing to the time window [0ms; 400ms] and re-applying the method, plateaus and edges in the same time windows occur, while subtle edges in results from the larger time window are more pronounced.

The difference in absolute values originate mainly from the normalization of synchronization index. Figure 10b presents the results from surrogate data, i.e. the randomized time series. These synchronization indices show a poor temporal structure and much lower values. Here and in the following, the t-test gives p-values < 0.001 for all time windows and both experimental conditions. That is the synchronization indices $MSS(t)$ in Fig. 10 are significantly different from the synchronization values of the surrogates in Fig. 10b.

Figure 11 shows average synchronization results obtained by averages over subsets of trials $0-19$, $20-39$ and $40-59$ in two different time windows for both conditions. It turns out that highly synchronized segments in the data occur in similar time windows as in Fig 10, however slightly shifted, shortend or lengthend. This finding indicates latency shifts in single trials.

Now, we focus to the shorter time window [0ms; 200ms] and classify components by their latencies and spatial distributions. Figure 12 presents averages over results of all trials and average current source density(CSD) maps corresponding to the detected time segments. We identify the components N100 and P200 in both conditions. In addition, these results reveal a time shift of component N100 between both experimental conditions. Hence component N100 depends on the cognitive task, and thus reflects an endogeneous underlying process. This finding contrasts to the hypothesis that N100 is an exogeneous component, i.e. independant from the cognitive task. The origin of this shift maybe attention or emotional effects by the subject.

We point out that the detected components in trial subsets are identified by the latency shift and the duration only, while the standard analysis also considers the spatial activity distribution on the scalp as well. However, here we neglect the discussion of the spatial topology as it is very noisy and thus does not allow a clear classification. Future research shall apply spatial denoising of components and we refer the reader to forthcoming work.

3.4. Application to Empirical Circular Data

Finally, we examine MPS in the data. Since the phases are defined in a corresponding narrow frequency band, the spectral density $A(t)$ and the global circular variance $R(t)$ are computed to indicate frequency bands of functional relevance. Figure 13 presents spectral power and global phase synchronization averaged over all trials. It turns out that there is low spectral power beyond 15Hz, i.e. the so-called $\beta-$band, while increased global phase synchronization is present at 17Hz and 20Hz. In addition, both large power spectral density and global phase synchronization occurs in the $\vartheta-$band about $\nu = 6$Hz in the distronic condition and in the lower $\vartheta-$band about $\nu = 5$Hz in the non-distronic condition, respectively. Hence, the subsequent analysis focus to the frequency bands $\nu = 6 \pm 0.6$Hz and $\nu = 5 \pm 0.5$Hz in the time window [0ms; 400ms] according to the analysis in the previous section.

Figure 14 shows results averaged over all trials, which exhibit short periods of increased MPS at $\sim$ 40ms, $\sim$ 80ms and $\sim$ 130ms in the distronic condition. Further, MPS is strong in [240ms; 340ms] and even stronger after 340ms. In the non-distronic condition, the results reveal increased MPS from stimulus onset to $\sim$ 90ms, between 110ms and 185ms and between 190ms and 240ms. After a longer transition period, strong MPS emerges at 290ms and even stronger between 340ms and 400ms. Hence, the time segments of increased MPS

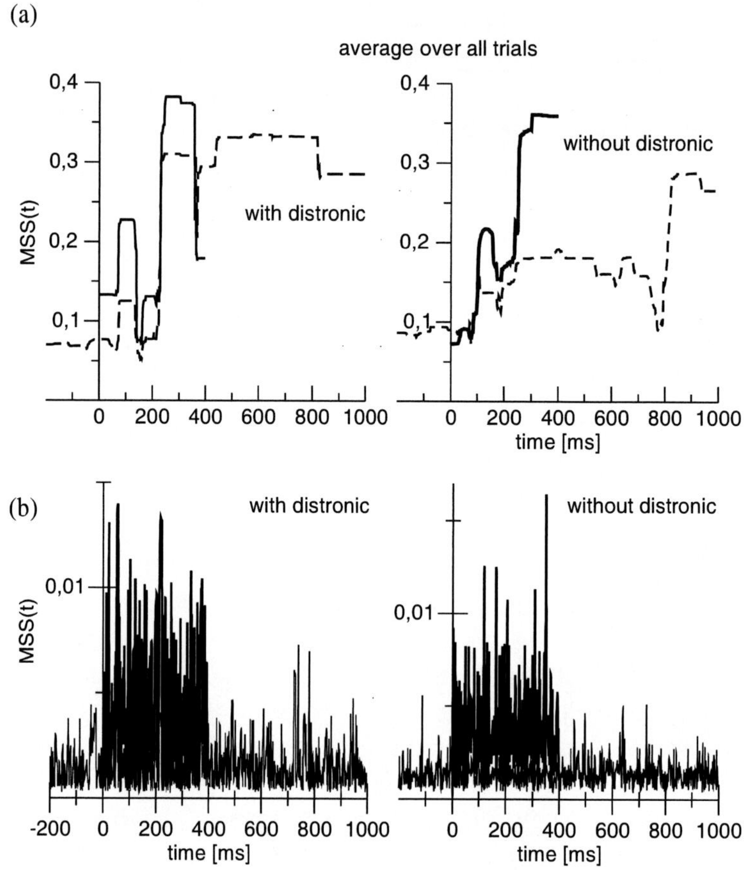

Figure 10. Cluster results for single averages over all trials. The cluster quality p quantifies the generalized synchronization $GS(t) = p(t)$ in time windows $[-200\text{ms}; 1000\text{ms}]$ (dashed line) and $[0\text{ms}; 400\text{ms}]$ for both experimental condition. The top part shows results from the original signal, where we observe a distinguished temporal structure. In contrast, the bottom part presents clustering results from the surrogate time-randomized data, which exhibits a poor temporal structure.

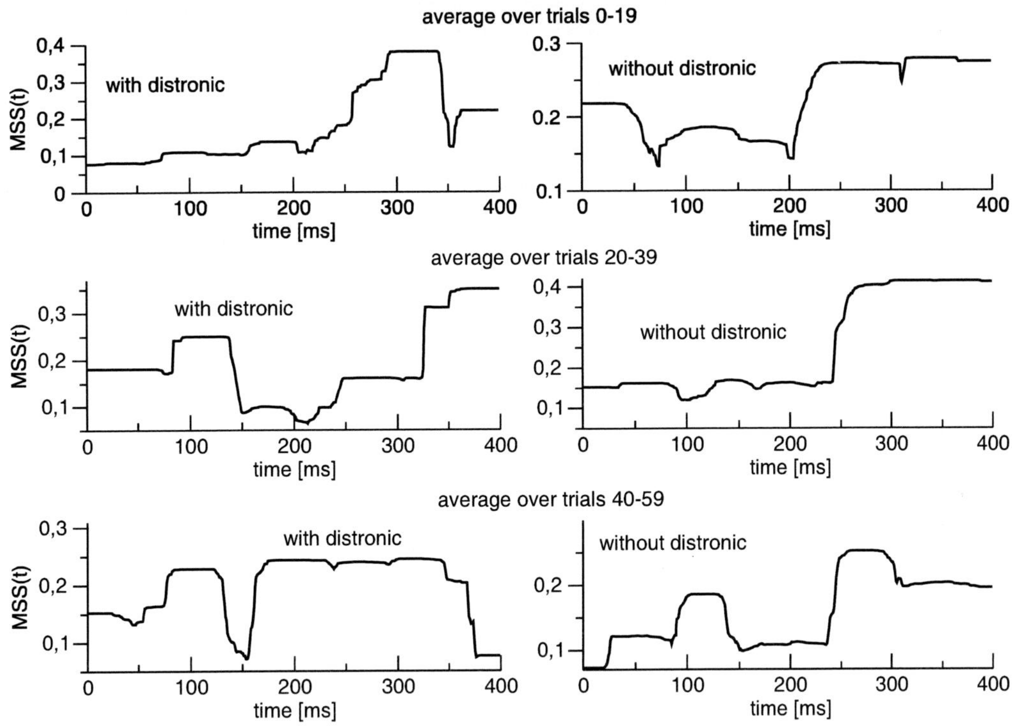

Figure 11. Cluster results from single data sets averaged over subsets of trials in two time windows for both experimental conditions. The cluster quality p quantifies the generalized synchronization $GS(t) = p(t)$ in time windows $[-200\text{ms}; 1000\text{ms}]$ (dashed line) and $[0\text{ms}; 400\text{ms}]$.

are different in both conditions, while strong MPS coincide after$\sim$ 240ms. Here and in the following, the t-test gives p-values < 0.001 for both experimental conditions, i.e. all results are statistically significant.

Since these results reflect the average behaviour of all trials and might be smeared due to latency shifts in single trials, the focus to averages of trial subsets to improve the temporal localization of segments. Figure 15 shows latency shifts between all subset averages at rather early latencies at about 90ms in the distronic condition. Further, all trial averages reveal a synchronous plateau of MPS about 130ms, while results from the averaged trials $0-19$ reveal retarded MPS at 260ms compared to the synchronous increase of MPS in subsets $20-39$ and $40-59$. This rather synchronous behaviour between different trials subsets does not occur in the non-distronic condition, where only the promiment plateau of subset $40-59$ at about 200ms coincides with the less prominent plateaus in $20-39$.

3.5. Comparison to an Existing Method

The global circular variance $R(t)$ is only a rough quantity for mutual phase synchronization, as it smears out spatial inhomogeneities by averaging. In contrast, the proposed phase synchronization index is based on a cluster algorithm in high-dimensional data space and thus

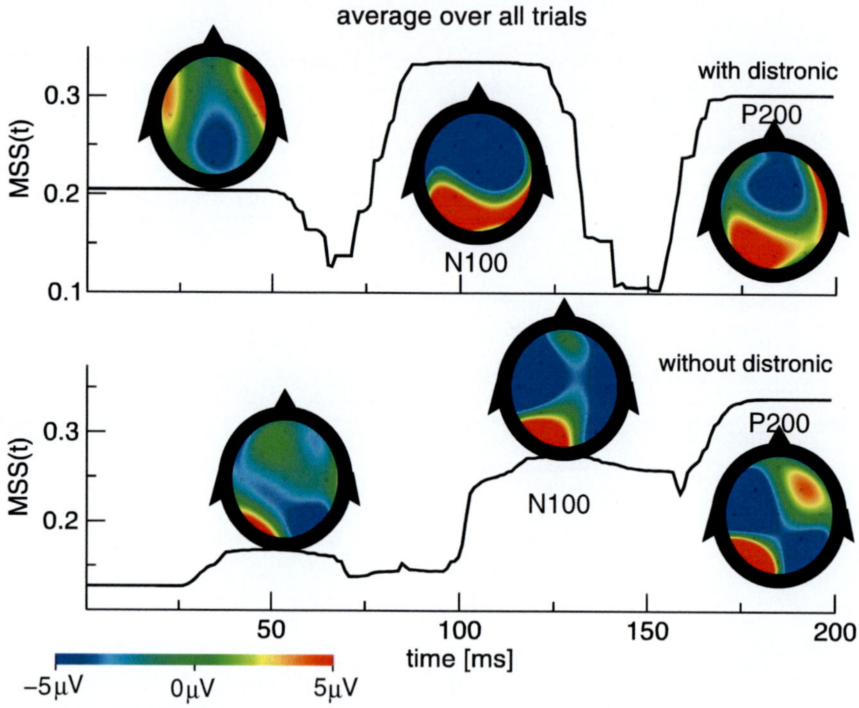

Figure 12. Cluster results from single data sets averaged over all trials for both experimental conditions and corresponding CSD-maps. The cluster quality p quantifies the generalized synchronization $GS(t) = p(t)$. Here, the focus to a shorter time window [0ms; 200ms] increases the analysis resolution and reveals clear temporal segments. The CSD-maps represent averages over the corresponding time intervals.

takes into account the space-time structure of data. Figure 16 presents a direct comparison of both quantities for results averaged over all trials. In the distronic condition, the circular variance behaves in time similar to the phase synchronization index. More detailed, transients at $\sim$ 170ms coincide in both quantities, while the transition from the synchronized time segment at 250ms to the time synchronized segments at 350ms occurs earlier in $R(t)$ than in $MPS(t)$. However, the most important difference between both quantities is the more detailed analysis of substructures by the proposed method. This is obvious in the non-distronic condition, where the substructure between 100ms and 250ms is lost in$R(t)$ and present in $MPS(t)$.

4. Discussion

The first part of the present work shows the relation of mutual space-time behaviour in brain signals to synchronization effects. We introduced the phrases mutual synchronization to describe the observed phenomena. The detected synchronization phenomena are transient and exhibit a drastical increase in the evolving time scale. This metastable behaviour in linear data is called mutual signal synchronization, while mutual metastability of circular

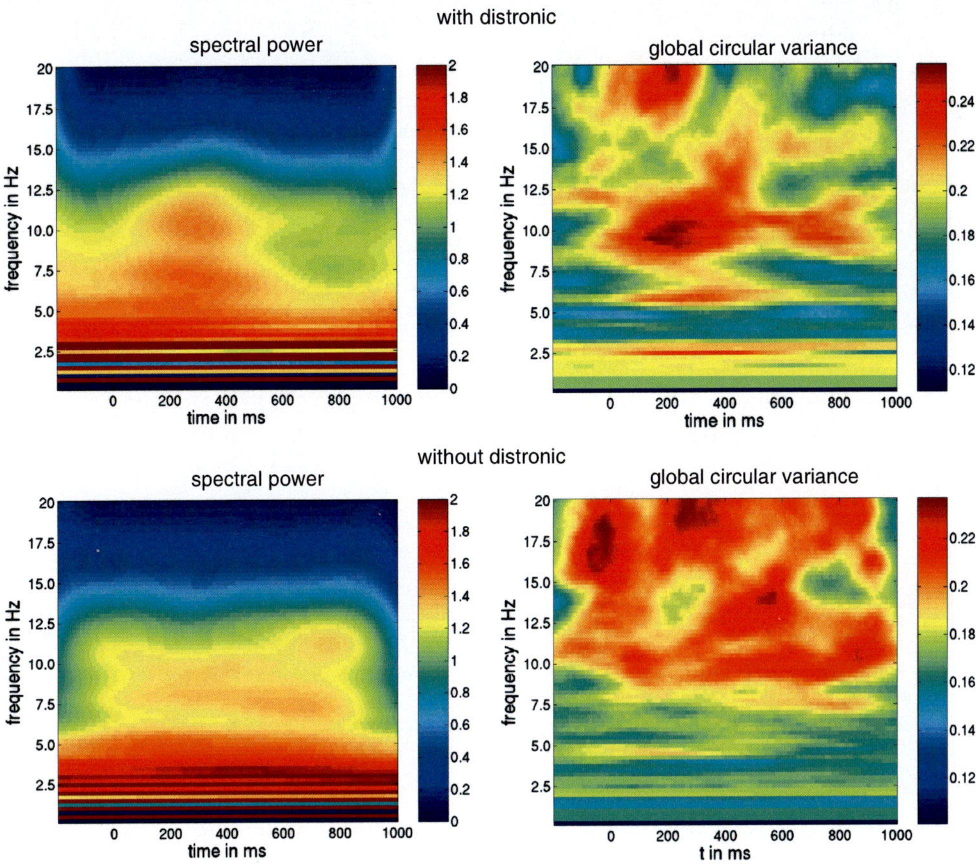

Figure 13. Spectral power from the average over all trials and global circular variance from all trials for both conditions. Spectral power contributions for frequencies larger than 20Hz are negligible.

data represent transient mutual phase synchronization. Considering these aspects, cluster analysis and the derived synchronization index p allows the segmentation of multivariate time series into metastable temporal segments. Brief applications to simulated linear and circular data illustrate the derived cluster method. The application to linear and circular evoked potentials led to highly synchronized time segments, which show good accordance to event-related components studied in neuropsychology. Investigating subsets of trials revealed latency jitters between the sets. These latency shifts indicate that external stimuli do not reset the phase of brain activity to the same value at the stimulus onset in each trial. Hence, our findings attenuate the general assumption of fixed time delayed evoked response to the stimulus onset similar to previous studies (see e.g. Pfurtscheller & da Silva (1999); Makeig et al. (2002); Tass (1999)). That is event-related potentials do not represent a linear superposition of signal and uncorrelated noise and, subsequently, single trial averages have to be interpreted cautiously. In addition to the detection of latency jitters, we found a latency shift of component N100 between both experimental conditions in the averages over all trials. This novel result may indicate attention or emotional effects at an early stage

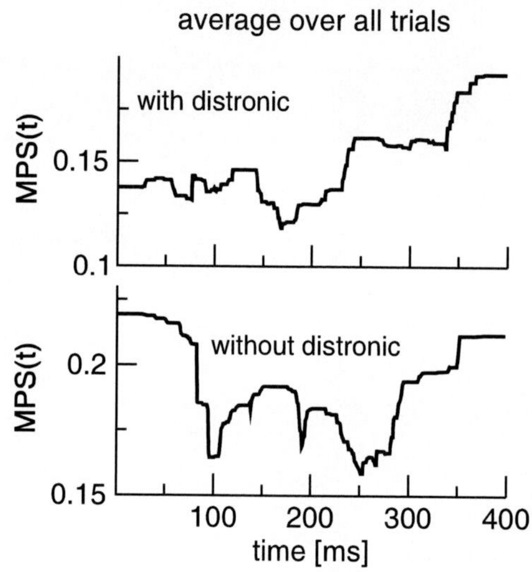

Figure 14. Cluster results from single phasic averages over all trials for both experimental conditions. The cluster quality p quantifies the mutual phase synchronization $MPS(t) = p(t)$. The original phasic signals are chosen in the freqency bands 6 ± 0.6Hz (with distronic) and 5 ± 0.5Hz (without distronic).

of cognitive processing.

The major reason for the successful analysis of averaged data from only few single data sets is the consideration of the space-time structure of the data. This becomes obvious by comparing our method to a conventional detection method for global phase synchronization neglecting spatial distributions. It turns out that the conventional method looses important temporal structures, which are extracted by the proposed approach. In future work, we aim to develop a thorough single trial analysis with improved statistical assessment in order to gain further insights to the phase synchronization processes of underlying neural activity.

References

Allefeld, C., Frisch, S., & Schlesewsky, M. [2005] "Detection of early cognitive processing by event-related phase synchronization analysis"*Neuroreport* **16**(1), 13–16.

Allefeld, C. & Kurths, J. [2003] "Multivariate phase synchronization analysis of eeg data" *IEICE Trans. Fundamentals* **E86-A**(9), 2218–2221.

Brandeis, D., Lehmann, D., Michel, C., & Mingrone, W. [1995] "Mapping event-related brain potential microstates to sentence endings"*Brain Topogrophy* **8**(2), 145–159.

Breakspear, M. [2002] "Nonlinear phase desynchronization in human electroencephalographic data" *Human Brain Mapping* **15**, 175–198.

Breakspear, M. & Friston, K. [2001] "Symmetries and itinerancy in nonlinear systems with many degrees of freedom" *Behavioral and Brain Sciences* **24**, 813–814.

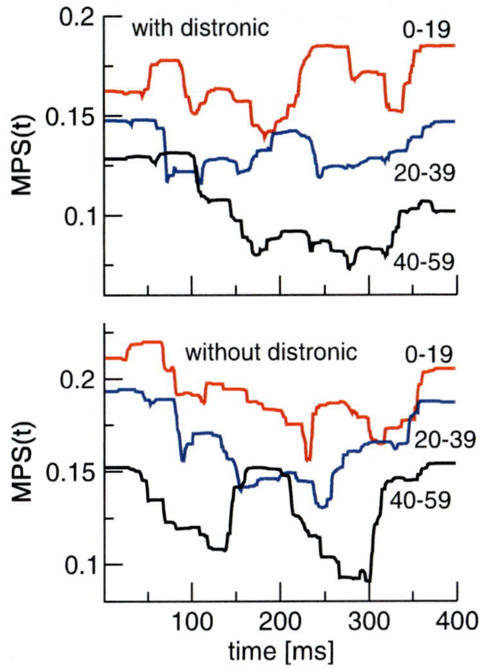

Figure 15. Cluster results from single phasic averages over subsets of trials for both experimental conditions. The cluster quality p quantifies the mutual phase synchronization $MPS(t) = p(t)$. Here, the phasic signals are chosen in the same freqency bands as in Fig. 14. For illustration reasons, results from $20-39$ and $40-59$ have been shifted artificially to lower values in both experimental conditions.

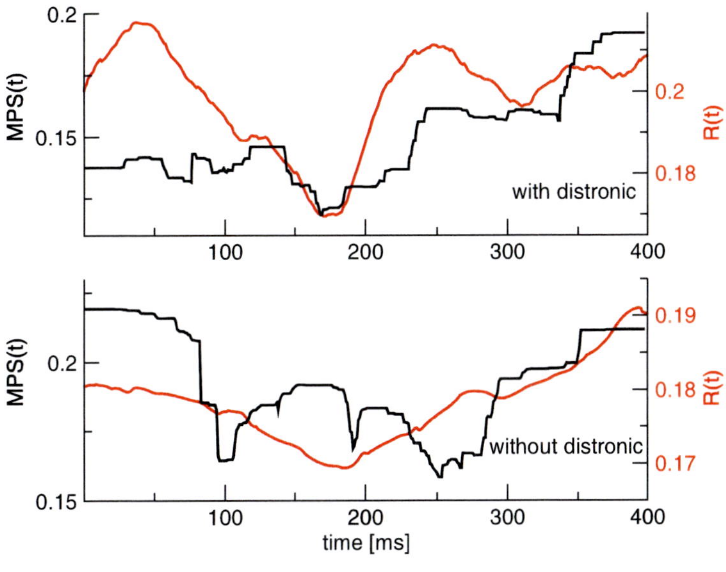

Figure 16. Comparison of cluster results to the conventional global circular variance for both experimental conditions.

Breakspear, M. & Terry, J. [2002] "Detection and description of nonlinear interdependence in normal multichannel human eeg" *Clin Neurophysiol* **113**, 735–753.

Busse, F. & Heikes, K. [1980] "Convection in a rotating layer: A simple case of turbulence" *Science* **208**, 173.

DeShazer, D., Breban, R., Ott, E., & Roy, R. [2001] "Detecting phase synchronization in a chaotic laser array" *Phys. Rev. Lett.* **87**(4), 044101.

Duda, R. & Hart, P. [1973] Pattern Classification and Scene Analysis. (Wiley, New York).

Freeman, W. [2000] Neurodynamics: An Exploration in Mesoscopic Brain Dynamics (Perspectives in Neural Computing). (Springer-Verlag, Berlin).

Freeman, W. [2003] "Evidence from human scalp eeg of global chaotic itinerancy" *Chaos* **13**(3), 1069.

Gratton, G., Coles, M., & Donchin, E. [1989] "A procedure for using multi-electrode information in the analysis of components of the event-related potential: Vector filter" *Psychophysiology* **26**, 222–232.

Haig, A., Gordon, E., Wright, J., Meares, R., & Bahramali, H. [2000] "Synchronous cortical gamma-band activity in task-relevant cognition" *Neuroreport* **11**, 669–675.

Hutt, A. [2004] "An analytical framework for modeling evoked and event-related potentials" *Int. J. Bif. Chaos* **14**(2), 653–666.

Hutt, A., Daffertshofer, A., & Steinmetz, U. [2003] "Detection of mutual phase synchronization in multivariate signals and application to phase ensembles and chaotic data" *Phys. Rev. E* **68**, 036219.

Hutt, A. & Kruggel, F. [2001] "Fixed point analysis: Dynamics of non-stationary spatiotemporal signals" in: S. Boccaletti, H. Mancini, W. Gonzales-Vias, J. Burguete, & D. Valladares, eds., *Space-time Chaos: Characterization, Control and Synchronization* pp. 29–44 (World Scientific, Singapore).

Hutt, A. & Riedel, H. [2003] "Analysis and modeling of quasi-stationary multivariate time series and their application to middle latency auditory evoked potentials" *Physica D* **177**, 203

Hutt, A., Svensen, M., Kruggel, F., & Friedrich, R. [2000] "Detection of fixed points in spatiotemporal signals by a clustering method" *Phys.Rev.E* **61**(5), R4691–R4693.

Ioannides, A., Kostopoulos, G., Laskaris, N., Liu, L., Shibata, T., Schellens, M., Poghosyan, V., & Khurshudyan, A. [2002] "Timing and connectivity in the human somatosensory cortex from single trial mass electrical activity" *Human Brain Mapping* **15**, 231–246.

Johnson, R. [1993] "On the neural generators of the p300 component of the event-related potential" *Psychophysiology* **30**, 90–97.

Karjalainen, P. & Kaipio, J. [1999] "Subspace regularization method for the single-trial estimation of evoked potentials" *IEEE Trans. Biomed. Eng.* **46**(7), 849–859.

Kay, L. [2003] "A challenge to chaotic itinerancy from brain dynamics" *Chaos* **13**(3), 1057–1066.

Kotz, S., Cappa, S., von Cramon, D., & Friederici, A. [2001] "Modulation of the lexical-semantic network by auditory semantic priming: An event-related functional mri study" *Neuroimage* **17**(4), 1761–1772.

Lachaux, J.-P., Lutz, A., Rudrauf, D., Cosmelli, D., Le Van Quyen, M., Martinerie, J., & Varela, F. [2002] "Estimating the time course of coherence between single-trial signals: an introduction to wavelet coherence" *Neurophysiol. Clin.* **32**, 157–174.

Lachaux, J.-P., Rodriguez, E., Martinerie, J., & Varela, F. [1999] "Measuring phase synchrony in brain signals" *Human Brain Mapping* **8**, 194–208.

Laskaris, N. & Ioannides, A. [2002] "Semantic geodesic maps: a unifying geometrical approach for studying the structure and dynamics of single trial evoked responses" *Clinical Neurophysiology* **113**, 1209–1226.

Lee, K., Williams, L., Breakspear, M., & E.Gordon [2003] "Synchronous gamma activity: a review and contribution to an integrative neuroscience model of schizophrenia" *Brain Research Reviews* **41**, 57–78.

Lehmann, D. & Skrandies, W. [1980] "Reference-free identification of components of checkerboard-evoked multichannel potential fields" *Electroenceph. Clin. Neurophysiol.* **48**, 609

Luecke, J. & von der Malsburg, C. [2004] "Rapid processing and unsupervised learning in a model of the cortical macrocolumn" *Neural Computation* **16**(3), 501–533.

Makeig, S., andT. P. Jung, M. W., andJ. Townsend, S. E., & andT.J. Sejnowski, E. C. [2002] "Dynamic brain sources of visual evoked responses" *Science* **295**, 690–694.

Mardia, K. & Jupp, P. [1999] Directional Statistics. (Wiley, New York).

Nunez, P. [1995] Neocortical dynamics and human EEG rhythms. (Oxfoord University Press, New York - Oxford).

Pascual-Marqui, R., Michel, C., & Lehmann, D. [1995] "Segmentation of brain electrical activity into microstates: Model estimation and validation" *IEEE Trans. Biomed. Eng.* **42**(7), 658–665.

Pfurtscheller, G. & da Silva, F. L. [1999] "Event-related eeg/meg synchronization and desynchronization: basic principles" *Clin Neurophysiol.* **110**(11), 1842–1857.

Pikovsky, A., Rosenblum, M., & Kurths, J. [2000] "Phase synchronization in regular and chaotic systems" *Int. J. Bif. Chaos* **10**(10), 2219.

Pikovsky, A., Rosenblum, M., & Kurths, J. [2001] Synchronization: A universal concept in nonlinear sciences. (Cambridge University Press).

Polich, J. [1998] "Clinical utility and control of variability" *J. Clin. Neurophysiol.* **15**(1), 14–33.

Rosenblum, M., Pikovsky, A., Schafer, C., Tass, P., & Kurths, J. [2000] "Phase synchronization: from theory to data analysis" in: F. Moss & S. Gielen, eds., *Handbook of Biological Physics* vol. 4 of *Neuroinformatics* pp. 279–321 (Elsevier, New York).

Rugg, M. & Coles, M. [1996] Electrophysiology of Mind. (Oxford University Press, Oxford).

Schirmer, A., Kotz, S., & Friederici, A. [2002] "Sex differentiates the role of emotional prosody during word processing" *Cognitive Brain Research* **14**(2), 228–233.

Schrauf, M. & Kincses, W. [2003] "Imaging the driver's workload using eeg/erp" in: *Vision in Vehicles* pp. 13–14 (Elsevier Science, Amsterdam).

Singer, W. & Gray, C. [1995] "Visual feature integration and the temporal correlation hypothesis" *Annual Review Neuroscience* **18**, 555–586.

Stam, C. & Dijk, B. [2002] "Synchronization likelihood: an un-biased measure of generalized synchronization in multivariate data sets" *Physica D* **163**, 236–251.

Tass, P. [1999] Phase resetting in medicine and biology : stochastic modelling and data analysis. (Springer, Berlin).

Tsuda, I. [2001] "Toward an interpretation of dynamic neural activity in terms of chaotic dynamical systems" *Behavioral and Brain Sciences* **24**(5), 793–847.

Uhl, C., Kruggel, F., Opitz, B., & von Cramon, D. Y. [1998] "A new concept for eeg/meg signal analysis: detection of interacting spatial modes" *Human Brain Map.* **6**, 137

Wackermann, J. [1999] "Towards a quantitative characterisation of functional states of the brain: From the non-linear methodology to the global linear description" *Int. J. Psychophysiology* **34**, 65–80.

Wilson, H. & Cowan, J. [1972] "Excitatory and inhibitory interactions in localized populations of model neurons" *Biophys. J.* **12**, 1–24.

Zaks, M., Park, E., Rosenblum, M., & Kurths, J. [1999] "Alternating locking rations in imperfect phase synchronisation" *Phys. Rev. Lett.* **82**, 4228.

In: Progress in Chaos Complexity Research
Editors: Franco F. Orsucci and Nicoletta Sala
ISBN: 978-1-60456-375-7

Chapter 2

A BRIEF NOTE ON RECURRENCE QUANTIFICATION ANALYSIS OF BIPOLAR DISORDER PERFORMED BY USING A VAN DER POL OSCILLATOR MODEL

Elio Conte[1,2], Antonio Federici[1], Gianpaolo Pierri[3], Leonardo Mendolicchio[3] and Joseph P. Zbilut[4]

[1] Department of Pharmacology and Human Physiology, University of Bari, 70100 - Bari, Italy
[2] T.I.R.E.S. - Center for Innovative Technologies for Signal Detection and Processing
[3] Department of Neurological and Psychiatric Sciences, Psychiatric Unit, University of Bari, 70100 - Bari, Italy
[4] Department of Molecular Biophysics and Physiology, RushUniversity, Chicago, IL60612,USA. E-mail: *fisio2@uniba.it*

ABSTRACT

Assuming a mathematical model based on van der Pol oscillator, we simulated the time course of the latent and acclaimed phases of the psychiatric pathology called bipolar disorder. Results were compatible with the analysis of experimental time series data of mood variation previously published by Gottschalk A. et al., (1995). Furthermore we performed Recurrence Quantification Analysis (RQA) of time series data generated by our mathematical model and we found that the obtained values for Recurrences, Determinism and Entropy may be considered as indexes of the increasing severity and stage of the pathology. We consequently suggest that these variables can be used to characterize the severity of the pathology in its observed stage. On the basis of the model, an attempt has also been made to discuss some aspects of the complex dynamics of the pathology. Results suggest that stochastic processes in mood variation of normal subjects play an important role to prevent mood from oscillating in a too rhythmically recurrent and deterministic way, as it occurs in bipolar disorder.

1. Introduction

Many attempts to understand the complex behavior of biological systems have been performed in recent years by applying the methodologies of non linear dynamics to the analysis of time series of biological data. This approach has been mainly applied to the study of heart rate variability, brain waves , enzymes and neurotransmitters (Degn H. et al., 1987).

Efforts have also been made to apply the principles of the sciences of complexity to relevant issues in psychology and psychiatry (Ehlers C.L., 1995). As a consequence a novel characterization of research methodology is arising in these disciplines. A classical conceptual framework in Psychology and Psychiatry is to look for stable differences between individuals and groups. The novel approach adopts the methodology of an intensive time sampling of multiple measures to reconstruct dynamic time patterns reflecting the dynamics of mental processes as they occurs on multiple occasions in different individuals. The vision is becoming relevant that abstract things as human feelings, mood and human behaviors do not change in time in a purely rhythmic and deterministic way, but their time variability may often reflect features of non linear dynamics and complexity. An increasing number of studies has investigated the process by which thoughts, feeling and behaviors of individuals unfold over time (Goldberger A. et al., 1990). In all of these studies repeated measures in time of considered variables have been employed to detect in the first instance the basic role of non linear dynamics in mental processes.

2. The Bipolar Disorder

One of the most investigated variables in Psychology and Psychiatry is represented by the size of time related changes in mood. Bipolar disorder, or manic depression, is a pathology related to mood fluctuations. Two primary forms of bipolar disorder do exist. The first one is characterized by a combination of manic and depressed episodes with the possibility of mixed episodes (Fawcett J. et al., 2000). The second one is characterized by a combination of hypomania and depressive episodes (Post R.M. and Luchenbaug D.A, 1992). Both are considered to be major mental pathologies that hit a valuable percent of adult population in the world. Psychiatrists follow fixed criteria for classification of bipolar disorder on the basis of the Diagnostic and Statistical Manual of Mental Disorders (American Psychiatric Association, 2000).

In 1995 a relevant paper was published by Gottschalk A. et al., which first evinced the possibility of chaotic mood fluctuations in bipolar disorder. On the basis of long term daily recordings of mood data, obtained from normal subjects and patients with bipolar disorder, the authors reconstructed the dynamic time pattern of mood in the two groups. Time series were recorded and analyzed over a time period of 2.5 years. The research was aimed to establish whether the temporal dynamics of recorded variables originates from a random or a periodic deterministic source. In order to evaluate the relevance of the results obtained by Gottschalk A. et al., we must consider that two basic models were classically referred to in studies on bipolar disorder before then. The so called Biological Rhythms Model was based on the observation of 48 hours mood cycles and on the tendency of mania periods to follow periods of depression. This model suggested an intrinsic periodicity for bipolar depression pathology (Wher T.A. et al., 1982). The so called Kindling Model was based on the observation that episodes in the pathology become more frequent as well as bipolar disorder progresses and that the later episodes appear to be spontaneous while earlier episodes seem to be induced by stressing agents (Post R.M. et al., 1985). The results obtained by Gottschalk A. et al. indicated that the mood in patients with bipolar disorder is not truly cyclic for extended periods of observation. In addition, the authors calculated correlation dimension of the recorded time series and obtained that mood dynamics in bipolar disorder can be described as a low-dimensional chaotic process, while the values of correlation dimension of the mood dynamics obtained in normal subjects were more similar to those characterizing random noise processes.

It is known that bipolar disorder become clearly acclaimed in humans after a time lag lasting up to several years, in which the pathology is mainly latent, with only moderate symptoms in mood changes over time. It is only in the following acclaimed phase of the pathology that an evident and quantifiable increase of bipolar mood changes clearly appears. In this work, by using a proper mathematical model, we have made an attempt to represent the qualitative changes in mood dynamics in a way which describe the transition from the latent to the acclaimed phase of bipolar disorder. Furthermore, by employing the method of the Recurrence Quantification Analysis (RQA; Webber C.L. and Zbilut J.P., 1994). A quantitative analysis of the most important variables which could characterize the transition from latent to acclaimed phase has been performed, by calculating the % of Recurrence, the % of Determinism and Entropy of time series data generated by the same mathematical model.

3. Qualitative Modeling of Bipolar Disorder by the Van Der Pol Mathematical Simulation

As above reported, Gottschalk A. et al. (1995) indicated that chaotic mood changes in bipolar disorder may exist. These authors considered their results also to explain the pathogenesis of the complex mood changes that characterize bipolar disorder. Although they did not observe linear periodic behavior in mood changes in their recorded time series data of subjects with bipolar disorder, they concluded that biological rhythms could still be involved in the pathogenesis of the complex mood variations in such a pathology. In this regard they evidenced that circadian rhythms supporting mood variations could be modeled by systems of differential equations as in particular by van der Pol oscillator equations. Also Daugherty D. et al. (2004) recently introduced mathematical models of bipolar disorder. They used limit cycle oscillators to model II kind bipolar disorder, by using van der Pol equations. It consequently seems that using van der Pol oscillator to model bipolar disorder can be of some interest to characterize some features of the dynamics of mood pathological changes.

This model, as well as almost any of the mathematical models employed in biology and medicine, can give only a very preliminary approach to bipolar disorder. It is unable to explain the intrinsic biological mechanism underlying the onset of the pathology but it can delineate a way of thinking about the dynamics of such a mechanism. As any preliminary mathematical model, its refinement and improvement will depend on the actual posses of daily recorded experimental time series of data describing the pathology. It is however built on some theoretical supports of intrinsic interest. To sketch the model we used a system of differential equations which, in its general form, can be written as it follows:

$$\frac{dx}{dy} = y\ ; \qquad (1.1)$$

$$\frac{dy}{dt} = A(1 - x^2)y - Bx - Cx^3 + D\mathrm{sen}\tau$$

$$\frac{d\tau}{dt} = E$$

In this model (x,y) are the mood state variables whose values characterize time dependent mood changes in bipolar disorder patients, as well as in normal subjects. A, B, C, D, E are parameters of the model. The values assigned to these parameters in our analysis were:

A=0.00001;
B = 5;
C=D=0;
$\Delta t = 0.0002$ arbitrary units.
As initial condition it was assumed x = 2 and y = 3.

In figure 1 (part a and part b) and figure 2 (part a and part b) an attempt to simulate time behavior of mood changes is reported, corresponding to the solutions x(t) and y(t) respectively. Such an attempt seems to work well to model the pathology transition from the latent to the acclaimed phase This point needs some further explanation. According to Gottschalk A. et al. (1995), also normal individuals show mood swings. A way to characterize the pathology is to designate how severe mood variation must be in amplitude. This is to say that only limit cycles with some minimal values of amplitude for x(t) and y(t) may be assumed to correspond to bipolar mood swings. On the basis of the results of Gottschalk A. et al. (1995) we assigned a value of about 6 as maximum value of amplitude for x(t) and y(t) when the pathology is still in the latent stage and a value of about 20 for acclaimed pathology. After assuming these values, figure 1 (part b) and figure 2 (part b) describe the progressive transition of the bipolar disorder from its latent to acclaimed stage as it is described by the model. It is evident that this approach looks at the disorder severity from the point of view of the amplitude of mood changes, which is only one of the several aspects involved in the progression of the pathology, but to deal with this only parameter is the aim of the present paper.

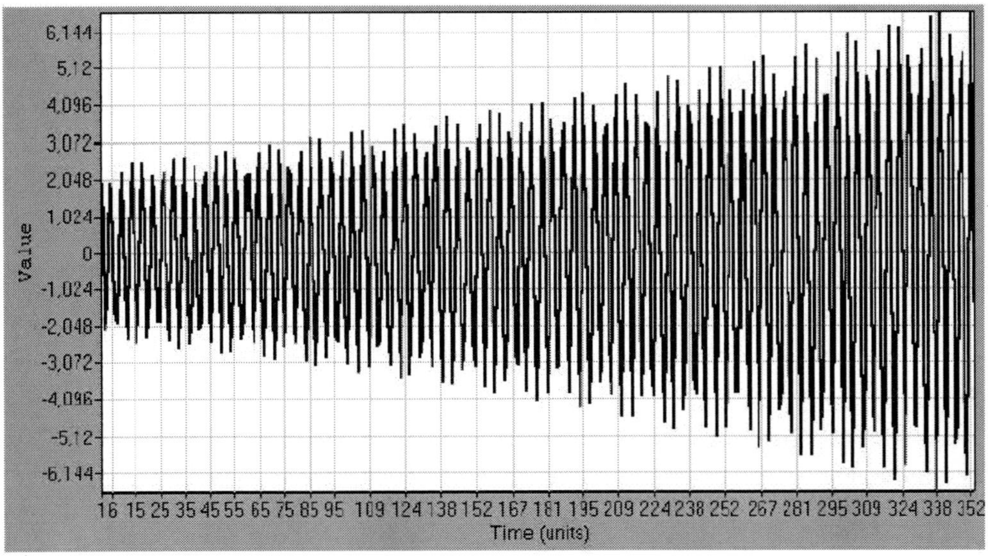

Figure 1A. A particular of the graph of the data of x(t) mood variation state in latent phase of bipolar disorder.

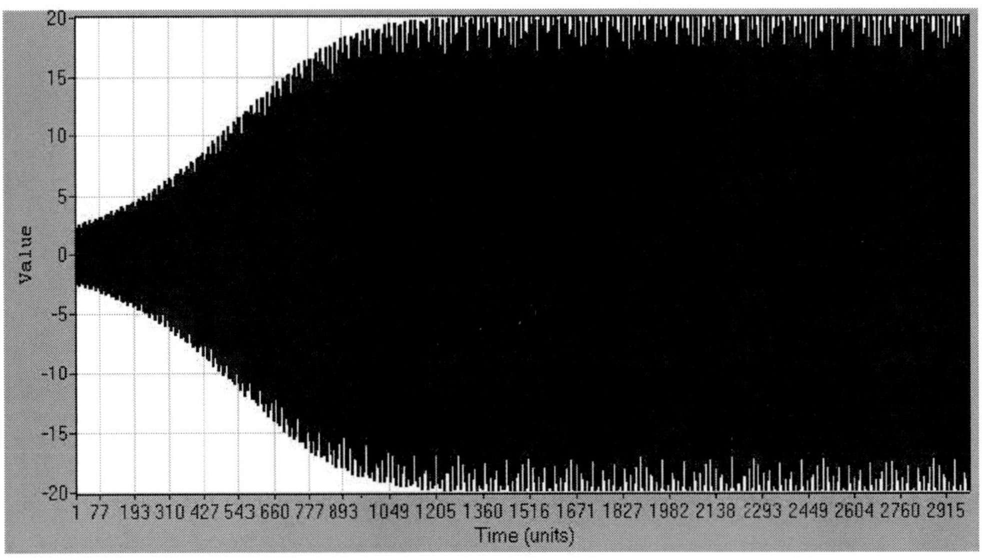

Figure 1B. Graph of the data of x(t) mood variation state in latent and acclaimed phases of bipolar disorder.

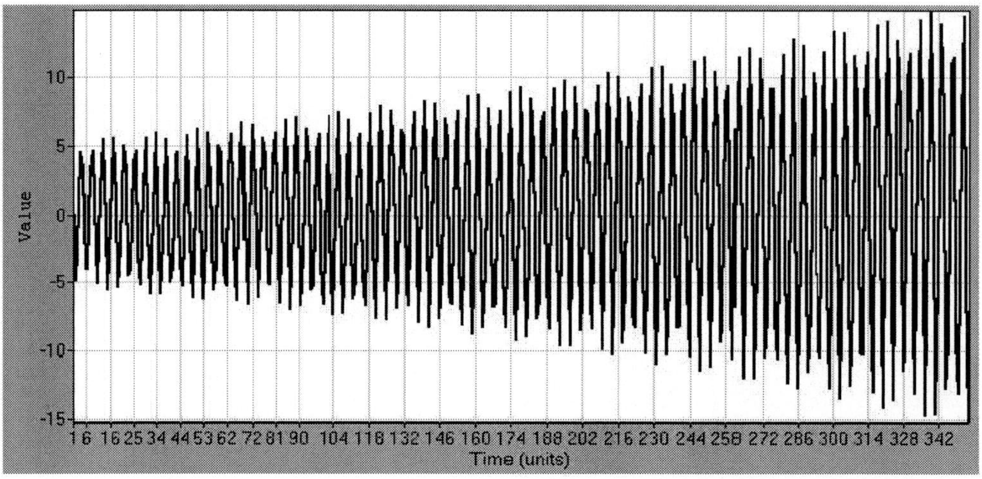

Figure 2A. A particular of the graph of the data of y(t) mood variation state in latent phase of bipolar disorder.

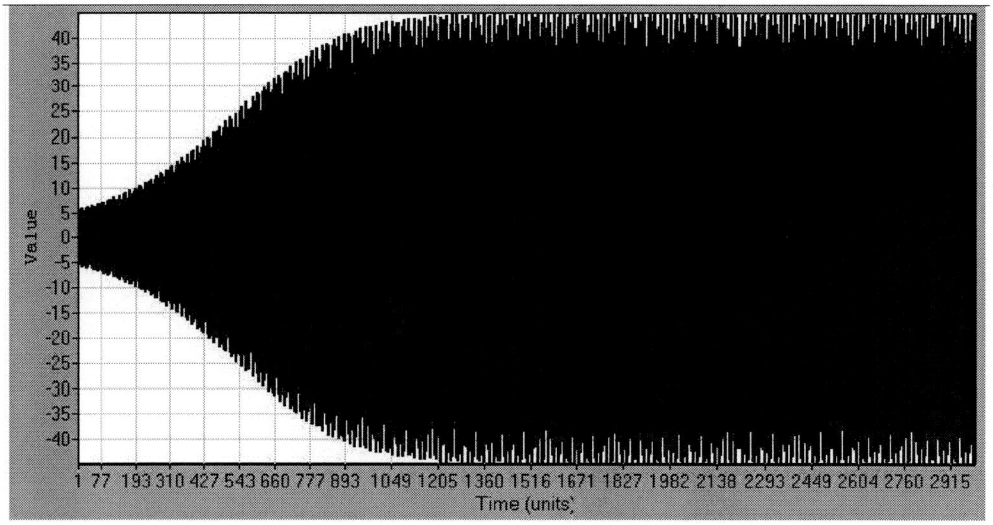

Figure 2B. Graph of the data of y(t) mood variation state in latent and acclaimed phases of bipolar disorder.

4. Quantitative Analysis of the Obtained Time Series Data

Gottschalk's Analysis of Correlation Dimension of Experimental Time Series

By using the method of Grassberger P. and Procaccia I. (1989), Gottschalk A. et al. (1995) obtained convergent estimates of the correlation dimension for six of the seven patients with bipolar disorder they studied. Five out of their seven patients gave values ranging from 1.1 to 3.8. The authors also calculated correlation dimension for the control subjects, but they did not obtain reliable convergent estimates of the correlation dimension in this group. It is worth noting that this was indeed a very interesting and intriguing result, as it will be discussed below. The authors concluded that the convergence of estimates for subjects with bipolar disorder gives important information on mood pathological dynamics, which in bipolar disorder show chaotic variations in time, and can be represented in a low dimensional phase space. This conclusion must be taken here with care and, as a first step, we should consider that the time series of mood variation that were recorded by Gottschalk A., did not satisfy the basic requirement of stationarity, that instead is fixed in order to apply some non linear methodologies as in particular the calculation of correlation dimension. To assess to what extent the conclusions by Gottschalk A. et al. (1995) may be or not invalidated by these limiting conditions, an attempt will be made in the next section to reproduce their results on the basis of the mathematical model we have introduced in the previous section.

Analysis of Correlation Dimension of Van Der Pol Model Generated Time Series

Our calculations were performed on two time series: the first one regarding only what we called the latent phase of the pathology (near 1000 points of the whole time series obtained by solving van der Pol equations given in 1.1); the second one including the whole time series (near 3000 points) simulating both the latent and the acclaimed phases of the pathology. Analysis of the data was made by first calculating Mutual Information and False Nearest Neighbors in order to reconstruct the phase space of data. Results are reported in figures 3 (part a and part b) and 4 (part a and part b). Analysis enabled us to conclude to use a time delay $\tau = 2$ and a phase space embedding dimension $d = 2$. The following step was to calculate correlation dimension respectively for the latent phase and for the whole behavior of the pathology. We also used surrogate data. The results are reported in figure 5 (part a and part b) and figure 6 (parts a, b, c and d) for the whole time series of latent plus acclaimed pathology. We obtained a convergent value of correlation dimension $D_2 = 1.26 \pm 0.0086$ in the case of x(t) mood state variable and $D_2 = 1.26 \pm 0.0118$ in the case of y(t) mood state variable as mean value for embedding dimension ranging from 5 to 10. Surrogate data gave respectively $D_2 = 5.54 \pm 0.83$ for x(t) and $D_2 = 5.54 \pm 0.82$ for y(t). Convergent values were obtained also in the case of analysis of the only latent phase of the pathology (1000 points time series) with $D_2 = 3.01 \pm 0.16$ in the case of x(t) and $D_2 = 2.97 \pm 0.20$ in the case of y(t). Surrogate data gave respectively $D_2 = 5.53 \pm 0.76$ for x(t) and $D_2 = 5.62 \pm 0.74$ for y(t). In conclusion, our model predicted chaotic mood variation in bipolar disorder in perfect accord with the previous results that were obtained by A. Gottschalk et al. (1995) In detail, our analysis predicted also a more increased chaotic behavior in what we called the latent phase of the pathology respect to the whole latent plus acclaimed phase.

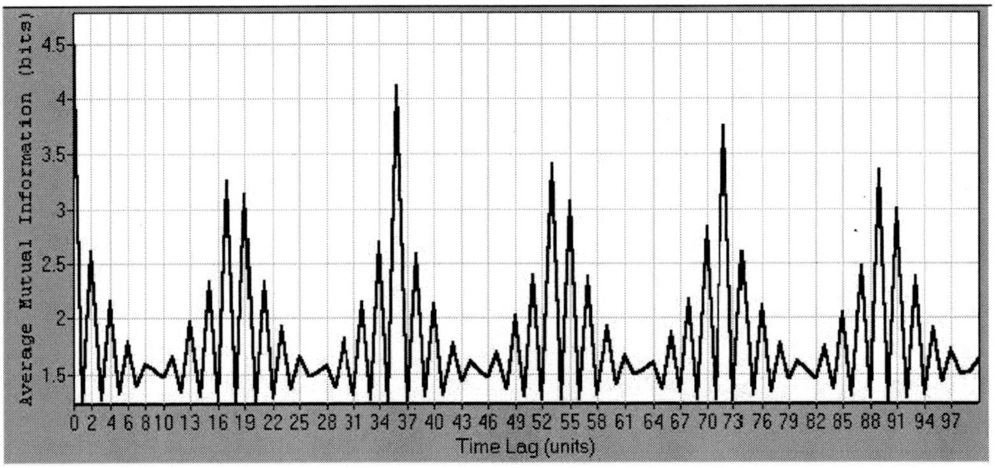

Figure 3A. Mutual Information of x(t) mood variation state.

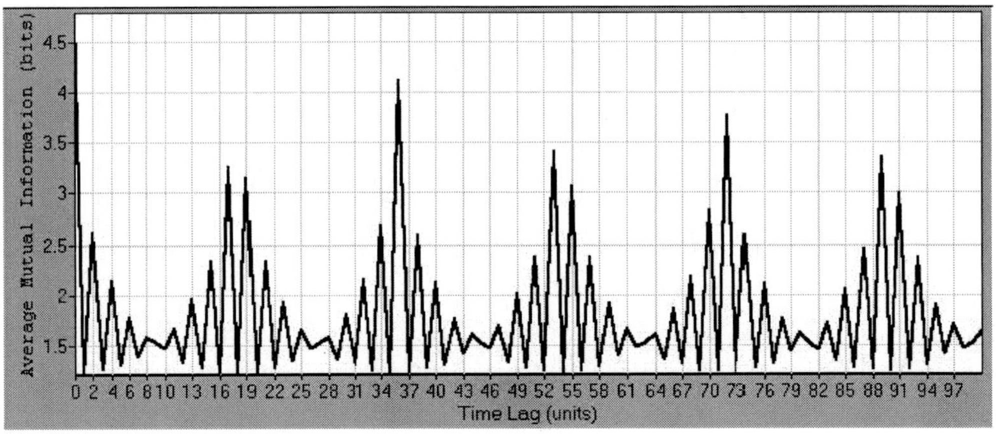

Figure 3B. Mutual Information of y(t) mood variation state.

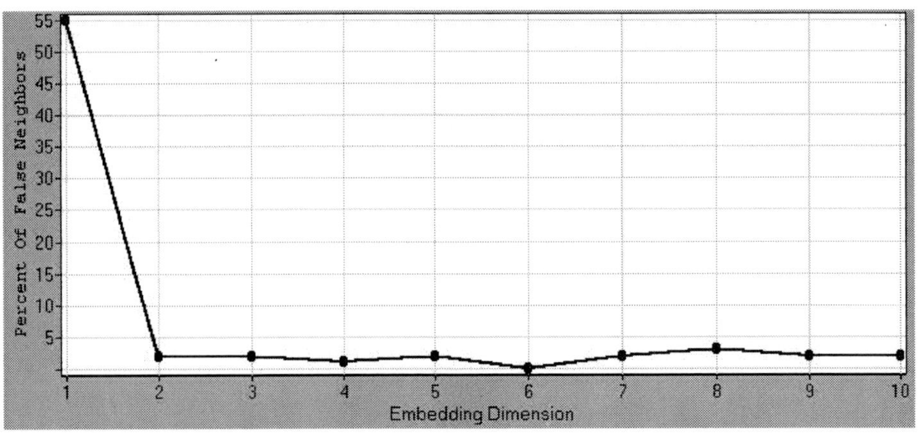

Figure 4A. False Nearest Neighbors of x(t) – data.

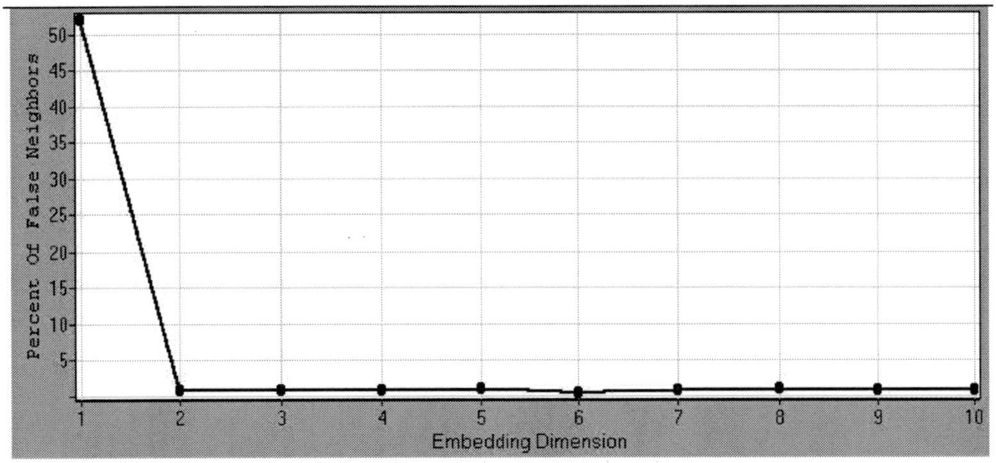

Figure 4B. False Nearest Neighbors of y(t) – data.

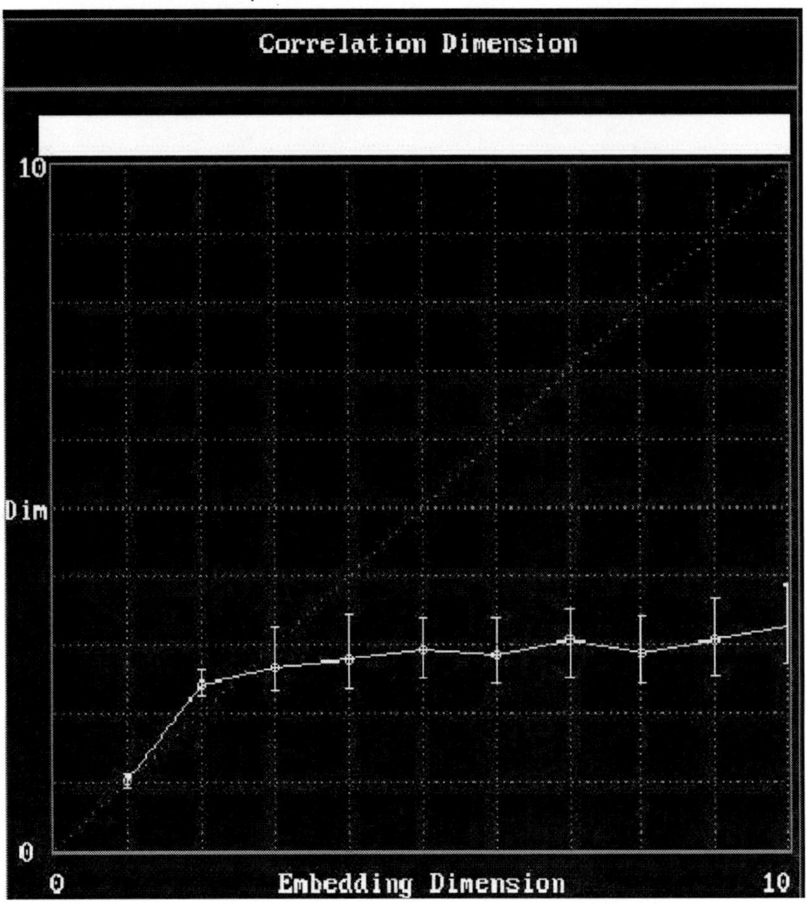

Figure 5A. Correlation Dimension of x(t) mood variation state in latent phase.

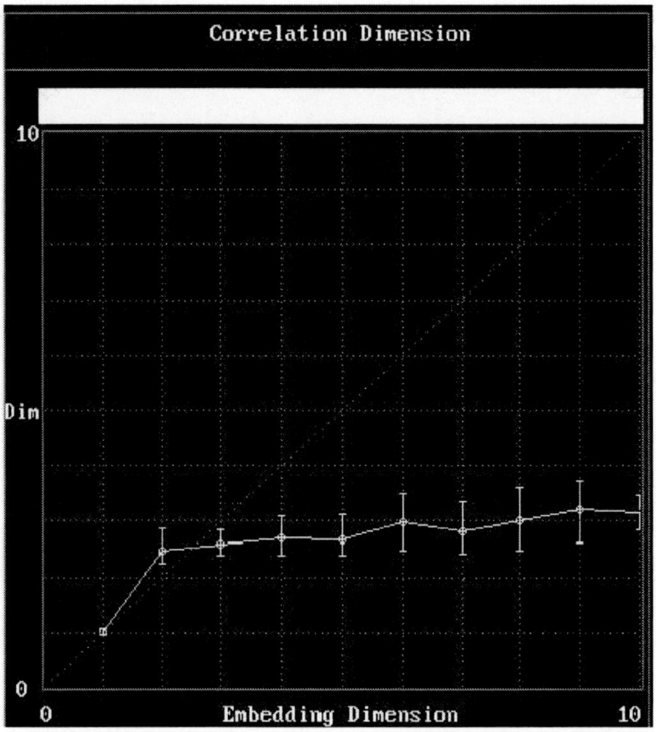

Figure 5B. Correlation Dimension of y(t) mood variarion state for latent phase of Bipolar disorder.

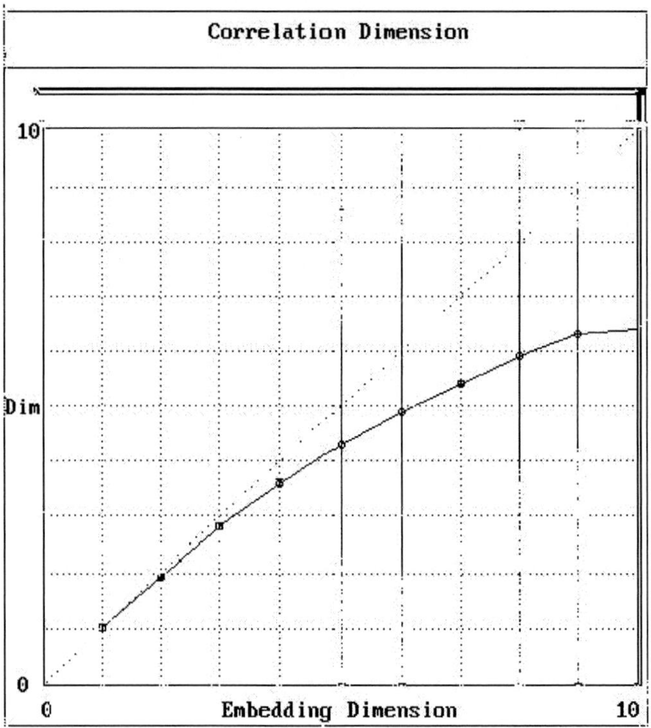

Figure 6A. Correlation Dimension x(t) surrogate data.

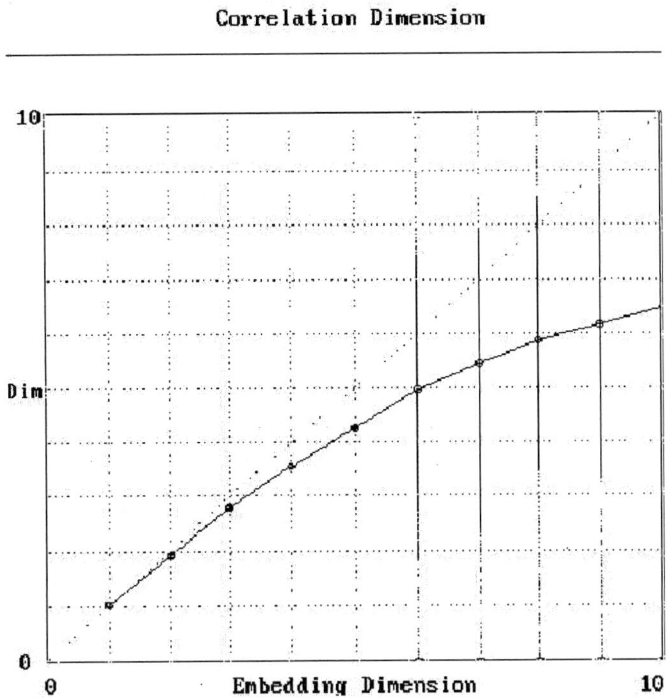

Figure 6B. Correlation Dimension y(t) surrogate data.

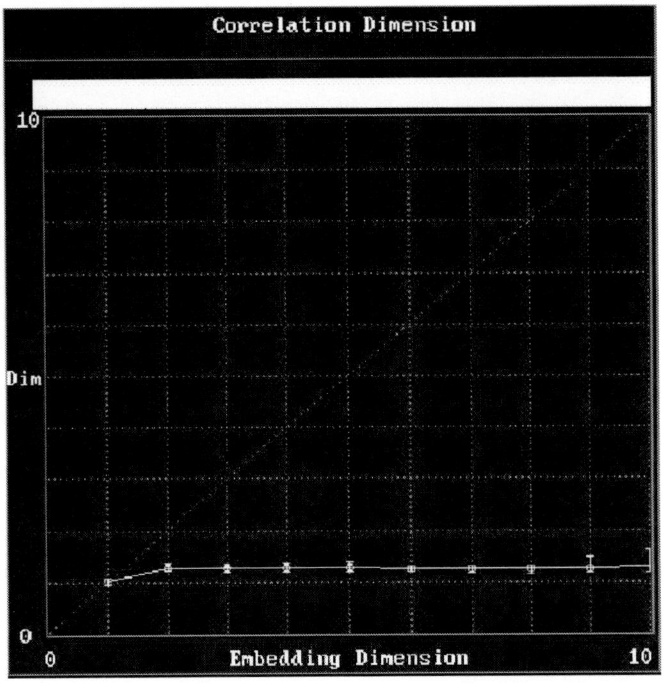

Figure 6C. Correlation Dimension of x(t) mood variation state in latent and acclaimed Phases of bipolar disorder.

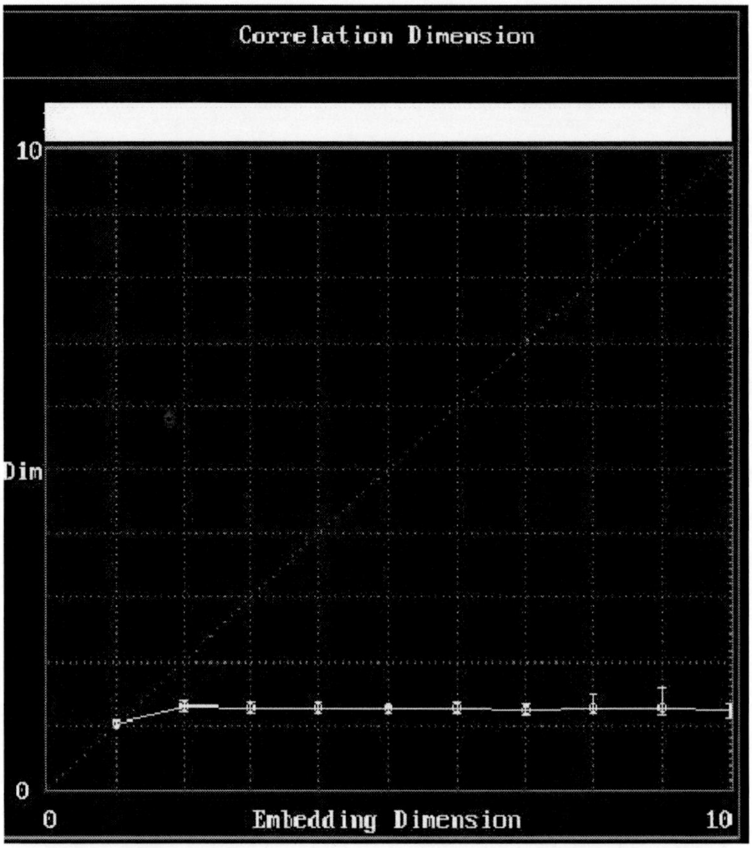

Figure 6D. Correlation Dimension of y(t) mood variation state in latent and acclaimed Phases of bipolar disorder.

To further proceed in the analogy between our model and the results of the quoted authors, we also tried to simulate mood variations of normal subjects. According to the results obtained by A. Gottschalk et al. (1995), we added a noisy component to x(t) and y(t) mood variation state variables in the time series of 1000 point regarding the latent phase. Actually we analyzed x(t) + n(t) and y(t) + n(t) with n(t) White Noise and Noise Amplitude ranging from 1 (3% of max amplitude of x(t) and/or y(t)) to 10, and mean = median = 0. In all of these analyses we obtained high dimensional phase space reconstruction. Reliable convergent estimates of the correlation dimension were no more obtained, in substantial accord with the Gottschalk's results. An example is reported in figure 7.

Lyapunov Exponents Calculation

By the calculation of the Lyapunov dominant exponent in the different cases of interest, we obtained: $\lambda_{dom}= 0.001 \pm 0.0009$ in the case of x(t) and y(t) for the whole phase of latent and acclaimed pathology; $\lambda_{dom} = 0.003 \pm 0.001$ for x(t) and $\lambda_{dom}=0.002 \pm 0.001$ for y(t) in the case of the only latent phase.

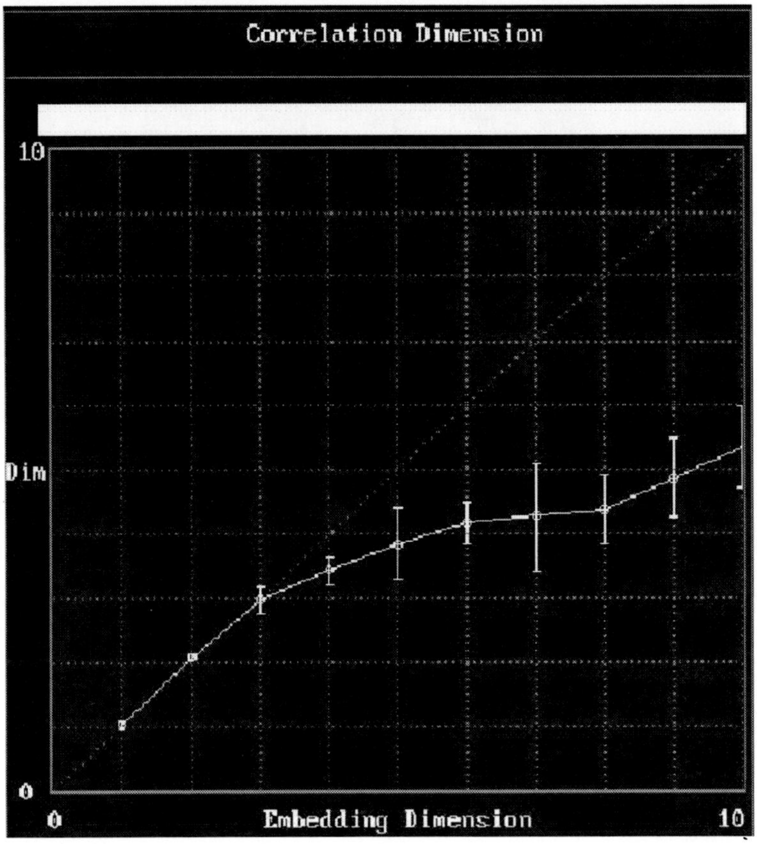

Figure 7. Correlation Dimension of x(t) mood variation state in presence of noise in Latent phase of bipolar disorder. Similar result for y(t).

Recurrence Quantification Analysis

Finally, we applied Recurrence Quantification Analysis (RQA) as it was introduced by J. Zbilut and C. Webber (Webber C.L., and Zbilut J.P.,1994). RQA is a very powerful method of non linear analysis since it enables to analyze also non stationary processes and it estimates the value of a number of variables that are relevant to characterize the basic features of the analyzed dynamic pattern. We concentrated our attention on three variables: the %Rec, the %Det. and the Entropy (E). %Rec quantifies periodicities in the time series, % Det evaluates determinism of the same time series and E relates the entropy. As parameters of RQA analysis we selected a time delay $\tau = 2$ and embedding dimension $d = 2$, the Euclidean radius $R = 4$ and the Distance $L = 4$. The results are reported in figure 8 (parts a, b, and c) and in figure 9 (parts a, b, and c) for x(t) and y(t), respectively. In the figure 8 (parts d and e) the recurrence plots are given for x(t) and y(t) mood variation state variable in the case of the whole dynamics, including both latent and acclaimed phases. It is evident that mood variation exhibits a rather chaotic behavior in the latent phase, but it shows more organization and structure in the acclaimed phase. Still, the determinism and periodicities in the acclaimed phase are well evident. Finally table 1 gives the values of %Rec., %Det. and E variables and their contributions in the two phases of the pathology simulation.

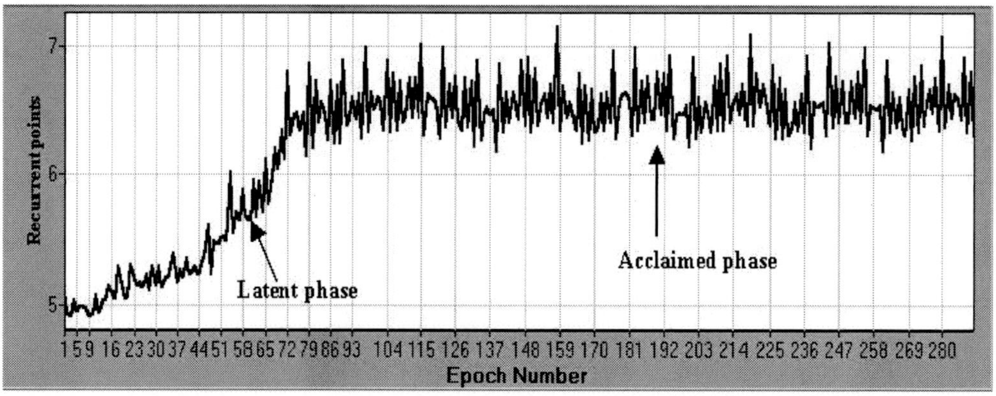

Figure 8A. % Recurrence of x(t) mood variation state in latent and acclaimed phase of bipolar disorder.

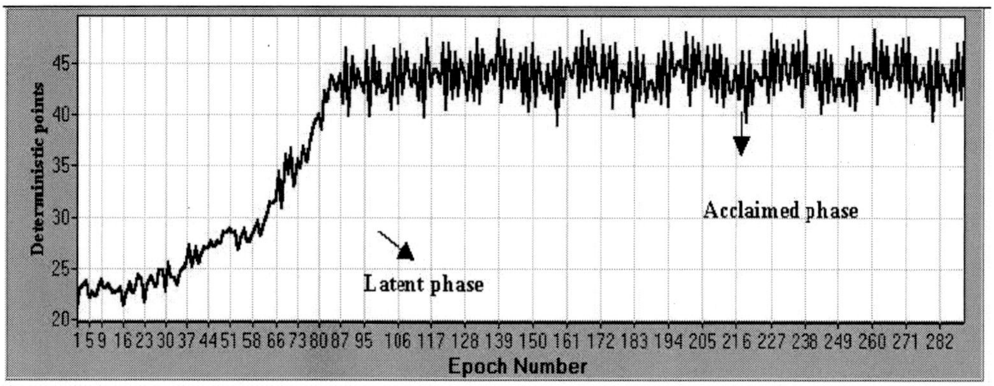

Figure 8B. % Determinism of x(t) mood variation state during latent and acclaimed phase of bipolar disorder.

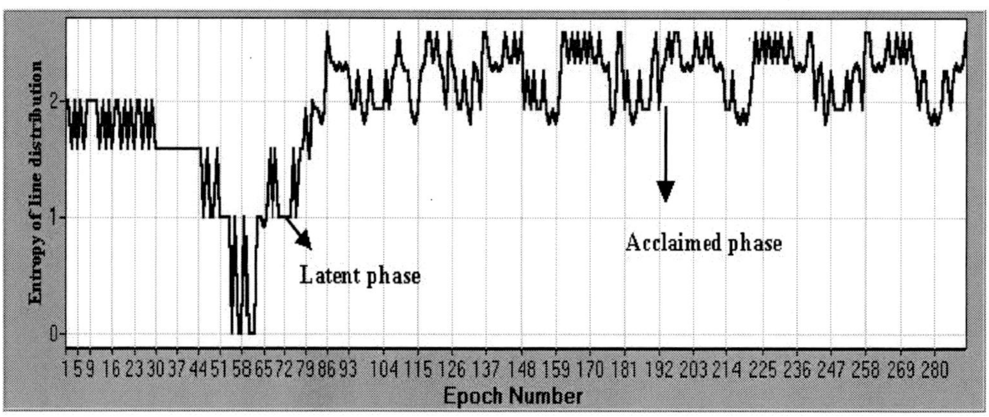

Figure 8C. Entropy of x(t) mood variation during latent and acclaimed phases of bipolar disorder.

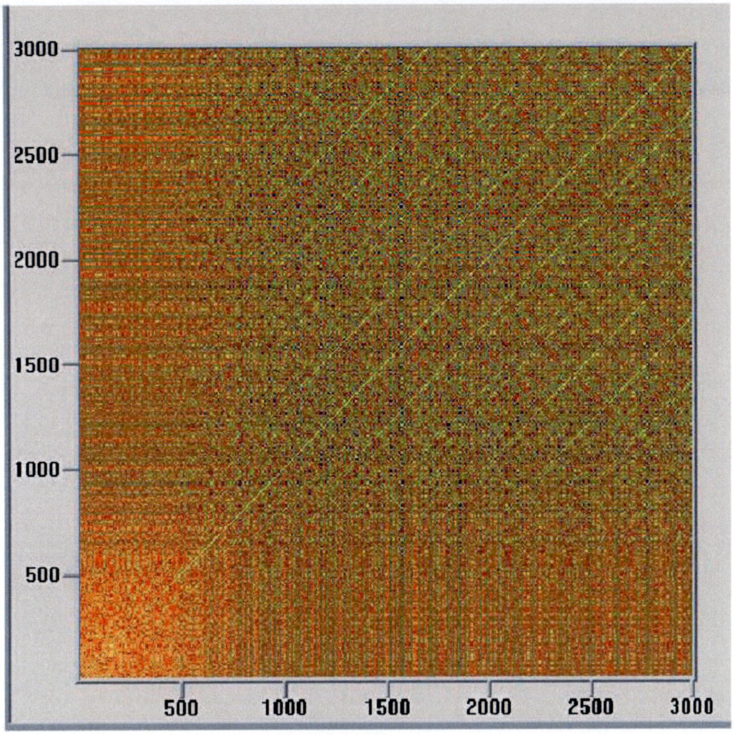

Figure 8D. Recurrence Plot of x(t) mood variation state during latent and acclaimed phases of bipolar disorder.

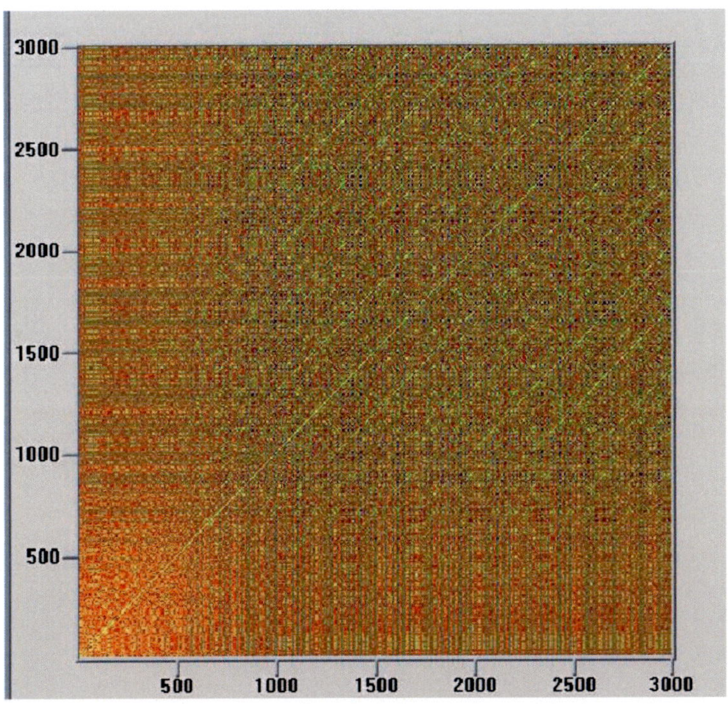

Figure 8E. Recurrence Plot of y(t) mood variation state during latent and acclaimed phases of bipolar disorder.

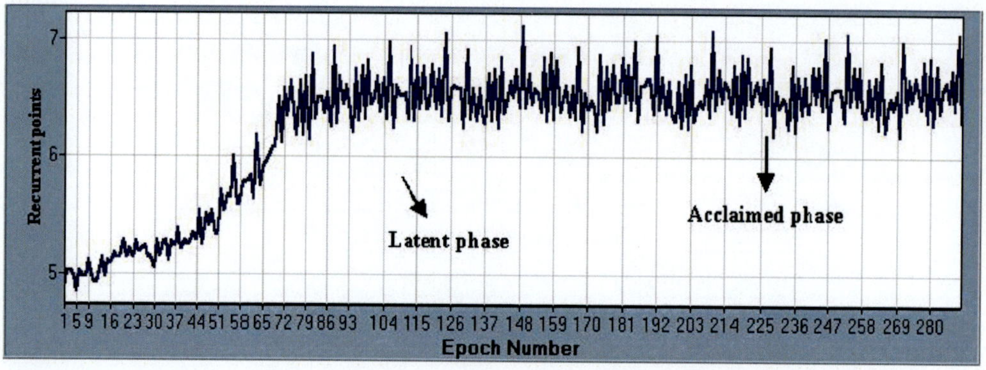

Figure 9A. % Recurrence of y(t) mood variation state in latent and acclaimed phases of bipolar disorder.

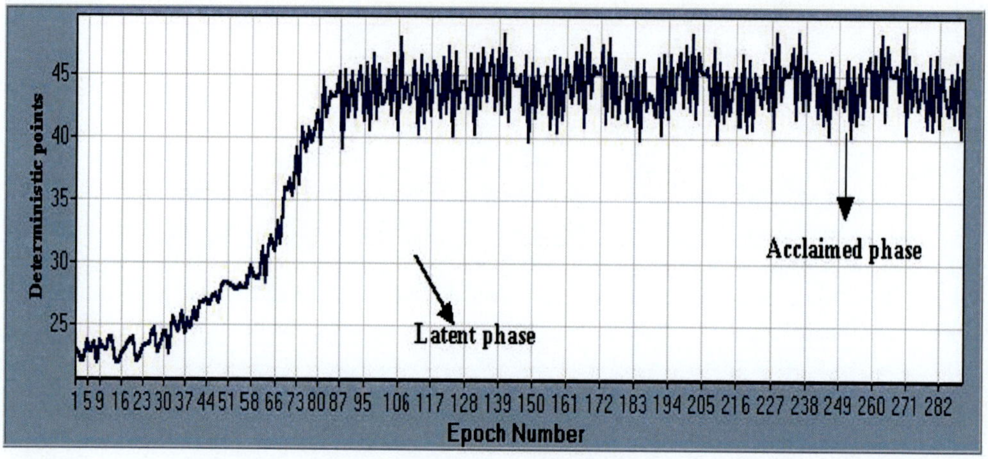

Figure 9B. % Determinism of y(t) mood variation state in latent and acclaimed phases of bipolar disorder.

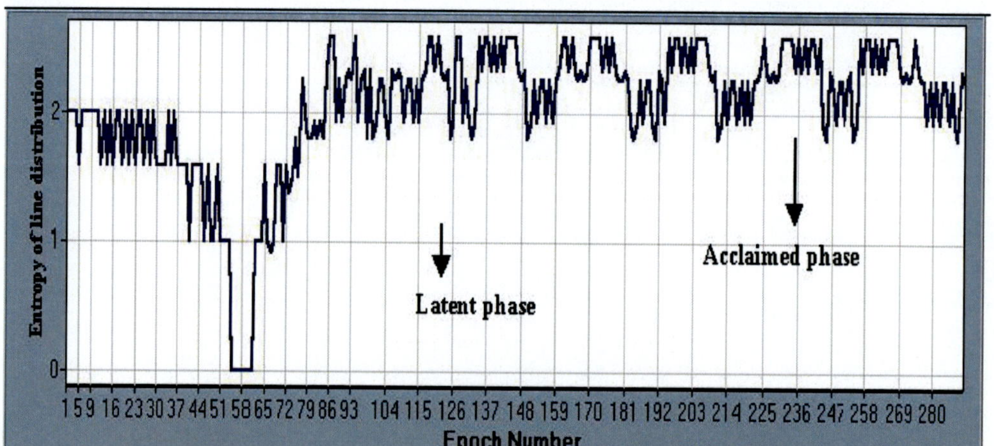

Figure 9C. Entropy of y(t) mood variation during latent and acclaimed phases of bipolar disorder.

Table 1. Mean values ± sd of %Recurrence (%Rec), %Determinism (%Det), and Entropy (E), calculated by Recurrence Quantification Analysis for x(t) and y(t) generated by van der Pol simulation of bipolar disorder, in the latent and acclaimed phase. The x(t) and y(t) signals reported in figure 1 and figure 2 have been divided into 71 epochs in the latent phase and 217 epochs in the acclaimed phase of the disorder. %Rec, %Det and E have been separately calculated for each epoch

	%Rec	%Det	E
x(t) Latent m ± sd; n = 71	5.389 ± 0.340	26.471 ± 3.891	1.501 ± 0.476
x(t) Acclaimed m ± sd; n = 217	6.542 ± 0.218	43.880 ± 2.517	2.247 ± 0.262
y(t) Latent m ± sd; n = 71	5.392 ± 0.391	26.283 ± 3.654	1.435 ± 0.593
y(t) Acclaimed m ± sd; n = 217	6.541 ± 0.220	43.898 ± 2.509	2.248 ± 0.271

Coarse Graining Spectral Analysis

To further test the results, we submitted time series data generated by our model also to the CGSA method (coarse graining spectral analysis), a methodology that was introduced by Yamamoto Y. And Hughson R.L. (1993) in order to quantify the true harmonic component and the true fractal component of a given spectrum in the frequency domain. By applying this method, we found that in the whole time series relating latent plus acclaimed phases of the pathology, the fractal power percent was 28.18% and harmonic power was 71.82% for x(t), with a fractal power percent of 15.39% and harmonic power percent of 84.61% for y(t). Percent fractal power increased up to 64.19% and harmonic power percent decreased to 35.21% in the case of x(t) mood variation in the latent phase, with percent fractal power of 58.91% and percent harmonic power of 41.09% for y(t).

5. Discussion

The results we obtained by analyzing time series data generated by the van der Pol model given in (1.1), seem not only to agree with the conclusions which were made by Gottschalk A. et al. in 1995 on the basis of true experimental time series data of mood variation in subjects affected by bipolar disorder, but they also suggest further considerations on the normal and pathological mood dynamics.

The quoted authors come to the relevant indication that mood variation of normal subjects seems to be mainly driven by random contributions, while the mood variation in subjects with bipolar disorder appears to me more organized and structured. This in our RQA analysis correspond to the occurrence of increasing values of recurrences, determinism and entropy when, in van der Pol mathematical simulation, the pathology enforces in time. This is a relevant technical point, because RQA enables us to study the non stationary time series generated by our mathematical simulation of Gottschalk's et al. (1995) experimental time series, while the calculation of correlation dimension performed by these authors should require stationary time series data, which is a condition not fully satisfied in their experimental approach. So their indicative conclusions are enforced by RQA analysis of simulated time series.

Another point to discuss is that we can simulate normal or pathological mood dynamics in our model by introducing or not white noise into it. According to Gottschalk A. et al. (1995) mood pathological dynamics can be represented in a low dimensional phase space and features chaotic variation in bipolar disorder. The possibility to represent the behavior of a process with a low number of dimensions in phase space is an intrinsic feature of low-dimensional attractors while high-dimensional phase spaces characterize random processes or deterministic processes with extraordinary complex dynamics. So the Gottschalk's et al. (1995) hypothesis was that mood variations of normal subjects should be determined by random processes, while the dynamics of mood variation in bipolar disorder should be driven by more deterministic processes. These processes are described by a finite and low number of dimensions in phase space and thus by an attracting pattern in a finite and low dimensional space. Therefore the dynamics of mood variation in bipolar disorder should exhibit more organization and more structure with respect to the case of normal subjects. If it is so, this should represent a result of considerable importance if we look at it in the more general framework of the decisive role of noise dependent dynamics in the normal and regular functioning of biological matter (as in the case of stochastic and stochastic-like resonance phenomena) and, in the present case, in the dynamics of mental variables. It is therefore a relevant result of this paper that, by introducing white noise into the van der Pol model, mood oscillations can be simulated which are similar to those of normal subjects. This agree with the hypothesis that, in contrast to the case of normal subjects in which mood variation seem to be regulated by random noise contributions, mood variation in bipolar disorder can occur only within more organized and structured patterns, in which recurrences and determinism become more dominant as the pathology progresses in time. If it is so it could be useful to diagnostic and to classify the stage of the pathology in a quantitative manner by estimation of the Recurrence Quantification Analysis variables as %Recurrences, %Determinism and Entropy.

Finally, the characterization of the dynamics of bipolar disorder as it emerges from the Recurrence Quantification Analysis of the behavior of the van der Pol oscillator, seems to provide some speculative insight on the functional characteristics of the mechanism that could be responsible for the advent of pathological mood variation in bipolar disorder. Physiological rhythms arise from largely non linear interactions between biological mechanisms and their fluctuating environment. A role is likely played by stochastic changes in these interactions to make them not too deterministic but as flexible and adaptive as it can be required by continuous changes in both the organism and its environment (Zbilut J.P. 2004). The hypothesis can be put forward that in bipolar disorder a drastic reduction could occur of the randomized phenomena which adaptively drive mood variation. As a consequence, internal rhythms of the neurobiological processes which govern mood would become too deterministic and poorly coupled with changes in the body internal and external environment.

This hypothesis about the mechanism for bipolar disorder should effort an unitary theoretical framework that puts in agreement psychological as well a neurobiological data (Extein I., 1979). To describe by a van der Pol mathematical model the influence of the environment, and the manner in which it may be correlated to the pathology of bipolar disorder, could be a next theoretical step.

REFERENCES

American Psychiatric Association, Diagnostic and Statistical Manual of Mental Disorders: DSM-IV-TR,Washington, 2000

Daugherty D., Roque-Urra T., Roque-Urra J., Snyderb J., Wirkus S., Porter M.A., Mathematical Models of Bipolar Disorders, arXiv:nlin.CD/0311032v2-16 Jun. 2004

Degn H., Holdan AV., Olsan L.F., Chaos in Biological Systems, New York Plenum Press,1987.

Ehlers C.L. Chaos and Complexity. Can it help us to understand mood and behavior? *Arch. Gen. Psychiatry.* 52, 960-964, 1995

Extein I., Potter W.Z., Wher T.A., Goodwin F.K., Rapid mood cycles after a nonadrenergic but not serotenergic antidepressant, *Am. J. Psychiatry,* 136, 1602-1603, 1979;

Fawcett J., Golden B., Rosenfeld N., New Hope for People with Bipolar Disorder, Prima Publishing, Roseville, 2000.

Goldberger A., Rigney D.R., West B.J., Chaos and Fractals in Human Psychology, *Sci. Am.* 262, 42-49, 1990.

Gottschalk A., Bauer M.S., Whybrow P.C., Evidence of Chaotic Mood Variation in Bipolar Disorder, *Arch. Gen. Psychiatry,* 52, 945-959, 1995

Grassberger P., Procaccia I., Measuring the Strangeness of Strange Attractors, *Physica.* D9, 189-208, 1989

Post R.M., David A. Luchenbaugh, Unique Design Issues in Clinical Trials of patients with bipolar Disorder, *Am. Journal of Psychiatry,* 149, 999-1010, 1992

Post R.M., Rublonow D.R., Ballenger J.C., Kindling Implications for the course of Effective Ilness, *Neurobiology of Mood Disorders,* 432-466, 1985, Baltimore.

Wher T.A., Goodwin F.K., Wirz-Justice A., Breltmaier J., Craig C., 48-hours sleep-wake cycles in manic-depressive illness: naturalistic observations and sleep deprivation experiments, *Arch. Gen. Psychiatry,* 39, 559-565, 1982

Wher TA., Sack D., Rosenthal N., Duncan W., Gillin J.C., Circadian Rhythm Disturbances in manic-depressive Illness, *Fed. Proc.* 42, 2809-2814, 1983, see also ref. 7 and references therein.

Webber CL., Zbilut J.P., Dynamical Assessment of Physiological Systems and States Using Recurrence Plot Strategies, *J. Appl. Physiol.* 76(2), 965-73, 1994.

Yamamoto Y., Hughson RL., Extracting Fractal Components from Time Series, *Physyca D,* 68, 250-264, 1993.

Zbilut J.P. Unstable Singularities and Randomness. Their importance in the Complexity of Physiological, Biological and Social Sciences. Elsevier, Amsterdam, 2004.

In: Progress in Chaos Complexity Research
Editors: Franco F. Orsucci and Nicoletta Sala
ISBN: 978-1-60456-375-7

Chapter 3

PARALLEL IMPLEMENTATION OF SHORTEST PATHS PROBLEM ON WEIGHTED INTERVAL AND CIRCULAR ARC GRAPHS

Pramod K. Mishra*
Dept. of Applied Mathematics; Birla Institute of Technology,
Mesra Ranchi-835215 (India); Dept. of Electrical and Electronics Engineering;
Indian Institute of Technology, Kharagpur; Kharagpur-India

ABSTRACT

We present an efficient parallel algorithm for the shortest path problem in weighted interval graphs for computing shortest paths on CREW PRAM which runs in $O(n)$ time with n intervals in a graph. We have given a linear processor CREW PRAM algorithm for determining the shortest paths in an interval graphs.

Keywords: Parallel Algorithms, Shortest Paths Problem, Weighted Interval Graphs.

1. INTRODUCTION

The single source shortest paths problem is that of computing a shortest path from a given "source" interval to all the other intervals. Our algorithm solves this shortest - paths problem on interval graphs optimally in $O(n)$ time, when we are given the model of such a graph, i.e. the actual weight. A minimum spanning tree of an edge -weighted graph is a spanning tree of the graph of minimum total edge weight. A shortest path tree rooted at a vertex r is a spanning tree such that, for any vertex ν, the distance between respected and ν is the same as in the graph.

* Pramod K. Mishra: Email: pkmishra@ieee.org, pkmishra@bitmesra.ac.in

Minimum spanning trees and shortest -path trees are fundamental structures in the study of graph algorithm for finding each are known.

Typically, the edge-weighted graph G represents a feasible network. Each vertex represent a site. The goal is to install links between pairs of site so that signals can be routed in the resulting network. Each edge of G represents a link that can be installed.

The cost of the edge reflects both the cost to install the link and the cost (e.g., time) for a signal to traverse the link once the link is installed. A minimum spanning tree represents the least costly set of links to install so that, for each site, the cost for a signal to be sent between the site and the root of the trees is small as possible.

The goal of a minimum spanning tree is minimum weight whereas the goal of a shortest path tree is to preserve distance from the root. The cost to install a set of links so that every site has a short path to the root is only slightly more than the cost just to connect all sites. The graph in which the weight of each edge is the Euclidean distance between the points. The weight of the shortest path tree is much more than the weight of a minimum spanning tree. Conversely, in the minimum spanning tree the distance between the root and one of the vertices is much larger than the corresponding shortest path distance. Nonetheless, there is a tree which nearly preserves distances from the root and yet weights only a little more than the minimum spanning tree.

The main result of this paper is that such trees exist in all graphs and can be found efficiently. Let G = (V,E) be a graph with none negative edge weights and a root vertex r. Let G have n vertices and m edges. Let $\omega(e)$ be the weight of edge $e \in E$. The distance $D_G(u, v)$ between vertices U and V in G is the minimum weight of any path in G between them. A node of an interval graphs corresponds to an interval and an edge is between two nodes in the graph iff the two intervals corresponding to these nodes intersect each other. Note that an interval or circular arc graph with *m*-nodes can have $O(n^2)$ edges. Our algorithm achieves the optimal $O(n)$ time bound by exploiting several geometric properties of this problem and by making use of the special *UNION-FIND* structure. One of the minimum weight circle - cover problem, whose definition we briefly review :

Given a set of weighted circular arcs on a circle, choose a minimum weight - subset of the circular arcs whose union covers the circles.

It is known that the minimum weight circle cover problem can be solved by solving q instances of the previously mentioned single source shortest-path problem, where q is the minimum number of arcs crossing any point on the circle.

It is the circle cover problem that has the main practical applications (Aho, Hopcroft, and Ullman, 1974) and the study of this shortest paths problem has mainly been for the purpose of solving the circle cover problem.(However, interval graphs and circular-arc graphs do arise in VLSI design, scheduling, biology, traffic control, and other application areas (Mishra and Sharma, 2002) so that our shortest- paths result may be useful in other optimization problems. More importantly, our approach holds the promise of shaving a $\log n$ factor from the time complexity of other problems on such graphs.

Note that, by using our single-source shortest-paths algorithm, the all pair shortest-paths problem on weighted -interval and circular arc graphs can be solved in $O(n^2)$ time, which is optimal. The previously best time bound for the all- pair shortest paths problem (Lee and Lee, 1984) on weighted interval graphs was $O(n^2 \log n)$.

An $O(n^2)$ time and space algorithm for the un weighted case of the all pair shortest path problem was given, and these bounds have been improved by Chen and Lee (Chen and Lee, 1994). We henceforth assume that the intervals are given sorted by their left end points and also sorted by their right end points. This is not a limiting assumption in the case of the main application of the shortest-paths problem, which is the minimum weight circle-cover problem. In the latter problem an $O(n \log n)$ preprocessing sorting step is cheap compared with the previously best bound for solving that problem which was $O(qn \log n)$ (by using q times the subroutine for solving the $O(n \log n)$ time each). Using our shortest-paths algorithm, the minimum weight circle cover problem is solved in $O(qn + n \log n)$ time where the $(n \log n)$ term is from the preprocessing sorting step when the sorted list of end points is not given as part of the input. Therefore in order to establish the bound we claim for the minimum- weight circle-cover problem, it suffices to give a linear time algorithm for the shortest- paths problem on interval graphs. The linear time solution to the shortest paths problem on interval graphs. Therefore, we mainly focus on the problem of solving in linear time, the shortest paths problem on interval graphs.

We also hence forth assume that we are computing the shortest paths from the source interval to only those intervals whose right end points are to the right of the right end points of the source; the same algorithm that solves this case can, of course be used to solve the case for the shortest paths to intervals whose left end points are to the left of the left end point of the source. Clearly we need not worry about paths to intervals whose right endpoints are covered by the source since the problem is trivial for these intervals-the length of the shortest path is simply the sum of the weight of the source plus the weight of the destination, provided the weights are all non - negative.

We consider the shortest-paths problem on interval graphs in which the weights of the intervals are non negative. The minimum-weight circle cover problem however does allow circular arcs to have negative weights. Bertossi (Bertossi, 1988) has already given a reduction of any minimum weight circle cover problem with both negative and non negative weights to one only non negative weights (to which the algorithm for computing shortest paths in interval graphs with non negative weights is applicable). Therefore it suffices to solve the shortest paths problem on interval graphs for the case of non negative weights. Bertossis (Bertossi, 1988) reduction introduces zero weight intervals, so it is important to be able to handle problems with zero weight intervals. We only show how to compute the lengths of shortest paths. Our algorithm can be easily modified to handle in $O(n)$ time and $O(n)$ space, the computation for actual shortest paths and a shortest path tree, i.e. a tree rooted at the source node such that the path in the tree from the root to each node of the tree is the shortest path in the graph between them. In the next section we introduce terminology needed in the rest of the section 3 and 4. Consider the special case of the shortest-paths problem on interval graphs with only positive weights. In particular, section 3 presents a preliminary sub optimal algorithm which illustrates our main idea and observation, and section 4 shows how to implement various computation steps of the preliminary algorithm so that it runs optimally in linear time.

Section 5 gives a linear time reduction that reduces the non-negative-weight case to the positive weight case, and it shows how to use the solution the shortest paths problem on interval graphs to obtain the solution to that on circular - arc graphs.

2. TERMINOLOGY

In this section we introduce some additional terminology. Given a weighted set S of n intervals on a line, a path from interval $I \in$ S to interval $J \in$ S is a sequence $\sigma = (J_1, J_2, \ldots, J_k)$ of intervals in S such that $J_1 = I$, J_k and J_i and J_{i+1} overlap for every $I \in \{1, 2, \ldots, k-1\}$. The length of σ is the sum of the weights of its interval s and σ is a shortest path from I to J if it has the smallest length among all possible I to J paths in S.

We say that an interval I contains another interval J

$$iff\ I \cap j = J\ . \tag{1}$$

We say that I overlaps with J *iff* there intersection is non empty, and that I properly overlaps with J *iff* they overlap but neither one contains the other.

An interval I is typically defined (Mishra and Sharma, 1997) by its two endpoints, i.e. $I = [a,b]$ where $a \leq b$ and a (resp. b) is called the left (resp. right) end point of I. A point x is to the left (resp. right) of interval

$$I = [a,b] \text{ iff } x < a \text{ (resp. } b < x \text{)} \tag{2}$$

We assume that the input set S consists of interval

$$I_1, \ldots I_n \text{ where } I_i = [a_i, b_i],\ b_1 \leq b_2 \leq \ldots \leq b_n\ , \tag{3}$$

and that the weight of each interval I_i is $\omega_i \geq 0$. To avoid unnecessarily cluttering the exposition, we assume that the interval have distinct end points, that is $i \neq j$ implies

$$a_i \neq a_j\ ,\ b_i \neq b_j\ ,\ a_i \neq b_j \text{ and } b_i \neq a_j \tag{4}$$

(the algorithm for non distinct end points is a trivial modification of the one we give).

Definition 1

We use S_i to denote the subset of S that consists of interval $I_1, I_2, \ldots, I_i$. We assume without loss of generality, that the union of all the I_i's in S covers the portion of the line from a_1 to b_n . We also assume, without loss of generality, that the source intervals is I_1.

Observe that for a set S* of intervals, the union of all the intervals in S * may form more than one connected component. If for two intervals I' and I'' in S*, I' and I'' respectively belong to two different connected components of the union of the intervals in S*, then there is no path between I' and I'' that uses only the intervals in S* .

3. Preliminary Algorithm

This section gives a preliminary $O(n \log \log n)$ time (hence sub optimal) algorithm for the special case of the shortest - paths problem on intervals with positive weights (Booth, and Luekher, 1976). This should be viewed as a " warm - up " for the next section, which gives an efficient implementation of some of the steps of this preliminary algorithm, resulting in the claimed linear - time bound. In section 5 we point out how the algorithm for positive weight intervals can also be used to solve problems with non - negative weight intervals.

We begin by introducing definitions that lead to the concept of an inactive intervals (Gupta, Lee and Leung , 1982) in a subset S_i , then proving Lemmas about it that are the foundation of the preliminary algorithm.

Definition 2

An extension of S_i, is a set S'_i, that consists of S_i, and one are more intervals (not necessarily in S) whose right end points are larger than b_i. (There are of course, infinitely many choices for such an S'_i).

Definition 3

An interval I_k in S_i ($k \leq i$) is inactive in S_i *iff* for every extension S'_i of S_i, the following holds:

Every $J \in S'_i - S_i$ for which there is an I_1 - to -J path in S'_i has no shortest I_1 - to -J path in S'_i that uses I_k.

An interval of S_i which is not inactive in S_i is said to be active in S_i.

Intuitively, I_k is inactive in S_i if the other intervals in S_i are such that, as for as any interval J with the right endpoint larger than b_i is concerned I_k is " useless" for computing a shortest I_1 - to -J path (In particular, this is true for $J \in \{ I_{i+1}, \ldots , I_n \}$) .

Lemma 1

The union of all the active intervals in S_i covers a contiguous portion of the line from a_1 to some b_j , where b_j is the right most endpoint of any active intervals in S_i .

Proof

If I_k, $k \leq i$, is active in S_i, then by definition there is a shortest I_1 - to - I_k path in S_i, implying that every constituent intervals of such a shortest I_1 - to - I_k path is active in S_i. It thus follows that every point on the contiguous portion of the line from a_1 - to - b_j where b_j is the right most endpoint of any active interval in S_i is contained in the union of all the active intervals in S_i.

The following corollary follows from Lemma 1.

Corollary 1

I_i is active in S_i, *iff* there is an I_1 - to - I_i path in S_i (i.e., if $\bigcup_{1 \leq k \leq i} I_k$ covers the portion of the line from a_1 - to - b_i).

Definition 4

Let $label_j\,(i)$, $j \leq i$, denote the length of a shortest I_1 - to - I_i path in S that does not use any I_k for which $k \geq j$.

By convention, if $j < i$, then $label_j\,(i) = +\infty$. (5)
Observe that , for all i, $label_1(i) \geq label_2(i) \geq, \ldots, label_n(i)$. (6)
For an $I_k \in S_i$ if there is no I_1 - to - I_k path in S_i, then obviously $label_i\,(j) = +\infty$, (7)
for every $j = k, k+1, \ldots, i$.

Lemma 2

If $i > k$ and $label_i\,(i) < label_i\,(k)$, then I_k is inactive in S_i . (8)

Proof

Since $label_i\,(i) < label_i\,(k)$, $label_i\,(i)$ is not $+\infty$. Hence there is an shortest I_1 - to - I_i path in S_i .Because $label_i\,(i) < label_i\,(k)$, it follows that there is an shortest I_1 - to - I_i path in S_i that does not use I_k : the union of the intervals on that I_1 - to - I_i path contains I_k (because $i > k$), and hence I_k is "Useless " for any $j \in S'_i - S_i$ where S'_i is an extension of S_i .

The following are immediate consequences of Lemma 2.

Corollary 2

Let $I_{j1}, I_{j2}, \ldots, I_{jk}$ be the active intervals in S_j , $j_1 < j_2 <, \ldots, < j_k \leq i$. (9)

Then $label_i(j_1) \leq label_i(j_2) \leq, \ldots, \leq label_i(j_k)$. (10)

Note that the right endpoints of the active intervals $I_{j1}, I_{j2}, \ldots, I_{jk}$ in S_i are in the same sorted order as that of their labels $label_i(j_1)$, $label_i(j_2)$, . . . , $label_i(j_k)$. Their left endpoints however, are not necessarily in such a sorted order.

Corollary 3

If I_i contains I_k (hence $i > k$) and $label_i(k) < label_i(i)$, then I_k is inactive in S_i. (11)

Lemma 3

If $i > k$ and $label_i(i) < label_{i-1}(k)$, then I_k is inactive in S_i.

Proof

That $label_i(i) < label_{i-1}(k)$ implies that $label_i(i)$ is not $+\infty$. Hence there is an I_1 - to - I_i path in S_i, and there is an I_1 - to - I_k path in S_i. There are two cases to consider.

(i) The shortest I_1 - to - I_k path in S_i does not need to use I_i. Then

$$label_{i-1}(k) = label_i(k), \tag{12}$$

and hence

$$label_i(i) < label_i(k). \tag{13}$$

By Lemma 2, I_k is inactive in S_i.

(ii) The shortest I_1 - to - I_k path in S_i does not use I_i. Then

$$label_i(k) \geq label_i(i) + \omega_k > label_i(i) \text{ (Since } \omega_k > 0\text{)}. \tag{14}$$

Again by Lemma 2, I_k is inactive in S_i.

Lemma 4

If interval I_k , $k > 1$, does not contain any b_j $(j < k)$ such that I_j is active in S_{k-1} , then I_k is inactive in S_i for every $i \geq k$.

Proof

It suffices to prove that I_k is inactive in S_k. Suppose that I_k is active in S_k. Then by Lemma 1, the union of all the active intervals in S_k covers the contiguous portion of the line from a_1 to b_k. (note that b_k is rightmost endpoint of any interval in S_k) .

This implies that I_k contains the right endpoint of at least one active intervals in S_k other than I_k . However, all the intervals in S_{k-1} ($= S_k - \{I_k\}$) that I_k intersects are inactive in S_{k-1}, and hence they remain inactive in S_k contradicting that I_k intersects some active intervals in S_k other than I_k .

We first give an overview of the algorithm. The algorithm scans the intervals in the order $I_1, I_2, \ldots, I_n$ (i.e. the scan is based on the increasing order of the sorted right endpoints of the intervals in S). When the scan reaches I_i , the following must hold before the scan can proceed to I_{i+1} :

(1) All the active intervals in S_i are stored in a binary search tree T .
(2) All the inactive intervals in S_i have been marked as such (Possibly at an earlier stage, when the scan was at some $I_{i'}$ with $i' < i$).
(3) If I_k $(k \leq i)$ is active in S_i , then the correct $label_i(k)$ is known.

If we can maintain the above invariants, then clearly when the scan terminates at I_n, we already know the desired $label_n$ (i)s for all I_i 's which are active in S_n. A post processing step will then compute, in linear time by CRCW parallel computational (Mishra, 2004) model (Concurrent Read Concurrent Write). The correct $label_n$ (i)s of the inactive I_i 's in S_n.

The details of the preliminary algorithm follow next. In this algorithm the right end points of the active intervals are maintained in the leaves of the tree structure T, one end point per leaf , in sorted order.

1- Initialize T to contain I_1 .
2- For i = 2, 3, . . . , n , do the following .
Perform a search in T for a_i .

This gives the smallest b_j in T that is $> a_i$. If no such b_j exists, then (by Lemma 4) mark I_i as being inactive and proceed to iteration $i+1$. So suppose such a , b_j exists.

Set $label_i(i) = label_{i-1}(j)+\omega_i$, and note that this implies that I_j remains active in S_i and has the same label as in S_{i-1} i.e. $label_i(j) = label_{i-1}(j)$. Next, insert I_i in T (of course b_i is then in the right most leaf of T). Then repeatedly check the leaf of I_k which is immediately to the left of the leaf for I_i in T, to see whether I_k is inactive S_i (by Lemma 3, i.e. check whether $label_{i-1}(k) < label\ l_i$, and , if I_k is inactive then mark it as such, delete it from T, and repeat with the leaf made adjacent to I_i by the deletion of I_k .Note that more than one leaf of T may be deleted in this fashion, but that the deletion process stops short of deleting I_j itself, because it is I_j that gave I_i its current label

$$(i.e.\ label_i\ (i) = label_{i-1}\ (j)+\omega_i \geq label_{i-1}\ (j)\ . \qquad (15)$$

Of course any I_l whose leaf T is not deleted is in fact active in S_i and already has the correct value of $label_i\ (l)$: it is simply the same as $label_{i-1}\ (l)$ and we need not explicitly update it (the fact that this updating is implicit is important, as we can not afford to go through all the leaves of T at the iteration for each *i.*).

When steps 2 terminates (at $i=n$), we have the values of the $label_n\ (l)$*'s* for the other intervals (those that are inactive in S_n).

(3). For every inactive I_i in S_n, find the smallest right endpoints $b_j > a_i$ such that I_j is active in S_n , and set $label_n\ (i) = label_n\ (j)+\omega_i$. Note that by Lemma 1, such an I_j exists and it intersects I_i . This steps can be easily implemented by a right -to- left scan of the sorted list of all the endpoints.

The correctness of this algorithm easily follows from the definitions lemmas and corollaries preceding it . Note that although a particular iteration in step 2 may result in many deletions from T , overall there are less than *n* such deletions .

The time complexity of this algorithm is $O(n \log n)$ if we implement T as 2-3 tree (Atallah and Chen, 1989), but $O(n \log \log n)$ if we use the data structure of (Gabow and Tarjan, 1985)(the latter would require normalizing all the $2n$ sorted endpoints so that they are integers between 1 and $2n$). The next section gives an $O(n)$ - time implementation of the above algorithm. Note that the main bottleneck is step, since the scan needed for step 3 obviously takes linear time.

4. A Linear Time Implementation

As observed earlier, the main bottleneck is step 2 of the preliminary algorithm given in the previous section. We implement essentially the same algorithm but without using the tree. Instead, we use a UNION -FIND structure (Gabow and Tarjan, 1985) where the elements of the sets are integers in { 1, . . . , n }, with integer i corresponding to interval I_i . Initially set i is $\{I\}$. (we often call a set whose name is integer i as set i , with the understanding that set i may contain other elements than i.) During the execution of step 2 we maintain the following data structures (Mishra and Sharma, 2002) and associated invariants (assume we are at index i in step 2):

(1) To each currently active interval I_j corresponds a set named j . If I_1 , I_2 , . . . , I_k , are the active intervals in S_i , $i_1 < i_2 <, \ldots, < i_k$ then for every $i_j \in \{ i_1 , i_2, \ldots , i_{k-1}\}$, the indices of the inactive intervals $\{ I_l / i_j < i_{j+1} \}$ are all in the set whose name is i_{j+1}. Set i_{j+1} , by definition consists of the indices of the above -mentioned inactive intervals, and also of the index i_{j+1} of the active interval I_{ij+1} . Note that since I_1 is always active, $i_1 = 1$ in the above discussion and the set whose name is 1 is a singleton (recall that a preprocessing step has eliminated intervals whose right endpoints are contained in interval (I_1) . The next invariant is about intervals that are inactive an and do not overlap with any active interval .

(2) Let Loose (S_i) denote the subset of the inactive intervals in S_i that do not overlap with any active interval in S_i . Observe that based on Lemma 1, every interval in Loose (S_i) is not empty, then let CC_1 , CC_2 ,. . . , CC_i , be the connected components of Loose (S_i) : There is a set named j_l for every such CC_l , where I_{jl}, is the right most interval in CC_l ,(I_{jl} , is interval in CC_l having the largest right endpoint) ; we say that such an inactive I_{jl} is special inactive. The μ (say) elements in set j_l correspond to the μ intervals in CC_l, more specifically, they are the contiguous subset of indices $\{ j_l -\mu+1, j_l -\mu+2, \ldots, j_l -1, j_l \}$. Note that $j_l -\mu$ is the set named $j_l -1$, if $1<l<t$ and that $J_t= i$.

(3) An auxiliary stack contains the active intervals $I_{j1}, I_{j2}, \ldots, I_{jl}$ mentioned in item (1) above, with I_{jk} at the top of the stack . We call it the active stack .

(4) Another auxiliary stack contains the special inactive intervals $I_{j1}, I_{j2}, \ldots, I_{jl}$ at the top of the stack. We call it the special inactive stack.

A crucial point is how to implement, in step 2, the search for b_j using a_i as the key for the search. This is closely tied to the way that the above invariants (1) - (4) are mentioned . It makes use of some preprocessing information that is described next.

Definition 5

For every I_i , let succ(I_i) be the smallest index l, such that $a_i <b_l$, i.e.

$$b_l= Min \{ b_r / I_r \in S , a_i <b_r \}. \qquad (16)$$

Note that $l \leq i$ and that $l= i$ occurs when I_i does not contain any b_r other than b_i. Also, observe that the definition of the Succ function is static (it does not depend on which intervals are active).

The *Succ* function can easily be pre computed in linear time by scanning right -to -left the sorted list of all the *2n* interval endpoints.

The significance of the *Succ* function is that, in step 2, instead of searching for b_j using a_i as the key for the search, we simply do a FIND ($Succ(I_i)$) :

Let j be the set name returned by this FIND operation. We distinguish three cases.

(1) $j=i$ then surely (I_i) does not overlap with any interval in S_{i-1} and it is inactive in S_i (by Lemma 4). We simply mark I_i as being special inactive push I_i on the special inactive stack , and move the scan of step 2 to index $i+1$.

(2) If $j <I$ and Ij is active in S_{i-1} , we set $label_i (l) = label_{i-1} (j)+\omega_i$. (17)

Then do the following updates on the two stacks .:

(a) We pop all the special inactive intervals I_{il} from their stack and , for each such I_{il} , we do *UNION* (i_l , i), which results in the disappearance of set i_l and the merging of its elements with set *I*, set *I* retains it old name.

(b) We repeatedly check whether the top of the active stack, I_{ik}, is going to become inactive in S_i because of I_i (that is, because $label_i (i) <label_{i-1} (i_k)$. If the outcome of the test is that I_{ik} becomes inactive, then we do *UNION* (i_k,i), *pop* I_{ik} from the active stack, and continue

with I_{ik-1}, etc.. If the outcome of the test is that I_{ik}, is active in S_i, then we keep it on the active stack, push I_i on the active stack, and move the scan of step 2 to index $i+1$.

If I_i is a active in S_i, $j=j_1$, and $label_i(l) < label_{i-1}(j_2)$ (18)

Then the sets $j_2, j_3, \ldots, j_k$ disappear and their contents get merged with set i.

(3) If $j < i$ and I_j is special inactive in S_{i-1}, then I_i does not overlap with any active interval in S_{i-1}, and it is inactive in S_i (by Lemma 4). However, I_i does overlap with one or more inactive interval I_j; more precisely, I_i overlaps with some connected components of *Loose* (S_{i-1}) whose rightmost intervals are contiguously stored in the stack of special inactive intervals. Let these connected components with which I_i overlaps be called, in left to right order $C_1, C_2, \ldots, C_h$. The right most intervals of C_1 is I_j. Let $I_{r2}, I_{r3}, \ldots, I_{rk}$ be the right most intervals of (respectively) $C_2, C_3, \ldots C_h$ (of course $I_{rh} = I_{i-1}$). Observe that the top h intervals in the special inactive stack are $Ij, I_{r2}, I_{r3}, \ldots, I_{rh}$, with I_{rh} $(=I_{i-1})$ on top. Because of I_i all of these h intervals will become inactive in S_i (whereas they were special inactive in i_{i-1}). Their h sets (corresponding to $C_1, C_2, \ldots, C_k$) must be merged into a new single set having I_i as its right most interval. I_i is special inactive in S_i. This is achieved by :

a) Popping $I_{rh}, \ldots, I_{r2}, I_j$ from the special inactive stack.
b) Performing *UNION* (r_h, i), *UNION* $(r_{h-1}, i), \ldots,$ *UNION* (r_2, i), *UNION*(j, i).
c) Pushing I_i on the special inactive stack.

Observe that the total number of the *UNION* and *FIND* operations performed by our algorithm is $O(n)$. It is well known (Booth and Luekher, 1976) that a sequence of m *UNION* and *FIND* operations on n elements can be performed in $O(m\alpha(m+n, n) + n)$ time (Chen and Lee, 1994), where $\alpha(m + n, n)$ is the (very slow growing) functional inverse of *ackermann's* function. Therefore, our algorithm runs within the same time bound. However, it is possible to achieve an $O(n)$ time performance for our algorithm by the following observations.

In our algorithm every *UNION* operation involves two set names that are adjacent in the sorted order of the currently existing set names (Mishra, 2004). That is, if L is the sorted list of the set names (Initially L consists of all the integers from 1 to n), then a *UNION* operation always involves two adjacent elements of L. Thus the underlying UNION - *FIND* structure we use satisfies the requirements of the static tree set in (Ibarrra, Wang and Zheng, 1992) inorder to result in linear - time performance : It is the linked list $LL = (1, 2, \ldots, n)$ where the element in LL that follows element is $next(l) = l+1$, for every $l = 1, 2, \ldots, n-1$ (the requirement in (Gupta, Lee, and Leung, 1982) is that the structure be a static tree). Note that the next function is static throughout our algorithm. The *UNION* operation in our algorithm is always of the form *unite* $(next(l), l)$, as defined (Golumbic, 1980), that is, it concatenates two disjoint but consecutive sub lists of LL into one contiguous sub list of LL. On this kind of structure a sequence of m *UNION* and *FIND* operations on n elements can be performed in $O(m+n)$ time (Golumbic, 1980). Therefore, the time complexity of our algorithm is $O(n)$.

Example: Interval Operations

We consider a set of n intervals $\mathbf{I} = \{I_1, I_2, \ldots, I_n\}$ of a line. (19)

Given an interval I, maxright (I) (*minright* (I)) denotes among all intervals that intersect the right endpoint of I , the one whose right end point is the farthest right(left) (see Fig 1). The normal definitions is as follows

$$\max right(I_i) = \begin{bmatrix} I_j & if \;\; b_j = \max\{b_k \mid a_k \leq b_i \leq b_k\} \\ nil & otherwise \end{bmatrix} \quad (20)$$

One way to compute the function *maxright* (and *minright* with the appropriate variations) is given in algorithm 1.

After step 1 of algorithm 1, we know that all the left endpoints of the intervals intersecting $I_i (1 \leq i \leq n)$ are on the left of its right endpoint b_i. Due to the definition of $d_i (1 \leq i \leq n)$ and of the prefix maximum on d_i at step 4, we are sure that for all the right endpoints $b_i (1 \leq i \leq n)$, e_i gives the right endpoint the furthest right of the intervals which intersect I_i and that num_i gives the number of the associated interval, that is to say *maxright* (I_i). We keep negative values for num_i $(1 \leq i \leq 2n)$ for left endpoints in order to be able at step 2 to distinguish the left endpoints from the right endpoints.

Step 1 requires $O(T_{S'}(n, p))$ time , whereas all the other steps require $O\left(\frac{n}{p}\right)$ local computations.

Step 1 and step 4 use a constant number of communications rounds. Then, *maxright* (and *minright* with the appropriate modifications) can be computed with time complexity $O(T_{S'}(n, p))$ and a constant number of communications rounds.

Algorithm 1: *Maxright*

Input: n intervals $I_i (1 \leq i \leq n) \left(\frac{n}{p} \text{ intervals on each processor}\right)$

Output: maxright (I_i) $(1 \leq i \leq n)$

Step 1: Global sort of the endpoints of the intervals in ascending order.

Step 2 : for each $i \in [1, 2n]$ do

$$\begin{bmatrix} assign\ to\ endpoint\ c_i\ the\ value\ d_i, defined\ by \\ d_i = \begin{cases} b_j & if \;\; c_i = a_j, \;\; for\ some\ 1 \leq j \leq n \\ 0 & if \;\; c_j = b_j, \;\; for\ some\ 1 \leq j \leq n \end{cases} \\ \end{bmatrix}$$

Step 3 : for each $i \in [1, 2n]$ do

$$\left[\begin{array}{l} assign\ to\ endpoint\ c_i\ the\ value\ num_i, defined\ by \\ num_i = \begin{cases} -j & if\ c_i = a_j,\ for\ some\ 1 \le j \le n \\ j & if\ c_j = b_j,\ for\ some\ 1 \le j \le n \end{cases} \end{array}\right]$$

Step 4 : Compute prefix maximum on the d_i, and let the result in $e_1, e_2, e_3, \ldots, e_{2n}$ on in the same time update the value num_i according to the following rule:

If $e_i \neq d_i$ and i > 1 set

$$\left[\begin{array}{lll} -\left|num_{(i-1)}\right|, & if\ c_i = a_j & for\ some\ 1 \le j \le n \\ \left|num_{(i-1)}\right|, & if\ c_i = b_j & for\ some\ 1 \le j \le n \end{array}\right]$$

Step 5 : for each $i \in [1, 2n]$ do

$$\left[\begin{array}{l} Set\ \max right(I_k) = I_j, \quad if\ c_i = b_k\ and\ num_i = j\ and\ k \neq j \\ Set\ \max right(I_k) = nil, \quad if\ c_i = b_k\ and\ num_i = j\ and\ k = j \end{array}\right]$$

Algorithm 2: ***Next***

Input: n intervals $I_i (1 \le i \le n) \left(\dfrac{n}{p}\ intervals\ on\ each\ processor \right)$

Output: *maxright* (I_i) $(1 \le i \le n)$

Step 1: Global sort of the endpoints of the intervals in ascending order.

Step 2: *for each* $i \in [1, 2n]$ *do*

$$\left[\begin{array}{l} assign\ to\ endpoint\ c_i\ the\ value\ d_i,\ defined\ by \\ d_i = \begin{cases} b_j & if\ c_i = a_j,\ for\ some\ 1 \le j \le n \\ 0 & if\ c_j = b_j,\ for\ some\ 1 \le j \le n \end{cases} \end{array}\right]$$

Step 3 : *for each* $i \in [1, 2n]$ *do*

$$\left[\begin{array}{l} \textit{assign to endpoint } c_i \textit{ the value } num_i, \textit{defined by} \\ num_i = \begin{cases} -j & \text{if } c_i = a_j, \text{ for some } 1 \le j \le n \\ j & \text{if } c_j = b_j, \text{ for some } 1 \le j \le n \end{cases} \end{array}\right]$$

Step 4 : Compute suffix maximum on the d_i, and let the result in $e_1, e_2, e_3, \ldots, e_{2n}$ on in the same time update the value num_i according to the following rule:

If $e_i \neq d_i$ and $i > 1$, set

$$\left[\begin{array}{lll} -\left|num_{(i-1)}\right|, & \text{if } c_i = a_j & \text{for some } 1 \le j \le n \\ \left|num_{(i-1)}\right|, & \text{if } c_i = b_j & \text{for some } 1 \le j \le n \end{array}\right]$$

Step 5 : *for each* $i \in [1, 2n]$ *do*

$$\left[\begin{array}{l} \textit{Set } next(I_k) = I_j, \quad \text{if } c_i = b_k \text{ and } num_i = j \text{ and } k \neq j \\ \textit{Set } next(I_k) = nil, \quad \text{if } c_i = b_k \text{ and } num_i = j \text{ and } k = j \end{array}\right]$$

We define also the parameter first *(I)* as the segment I which 'end first', that is , whose right endpoint is the furthest left (see Fig 1):

$$first\,(\mathtt{I}) = I_j, \quad with\ b_j = \min\{b_i \mid 1 \le i \le n\}.$$

To compute it, we need only to compute the minimum of the sequence of right endpoints of intervals in the family $\mathtt{I}$.

We will also use the function $next\,(I): \mathtt{I} \rightarrow \mathtt{I}$ defined as

$$next\,(I_i) = \left[\begin{array}{ll} I_j, & \text{if } b_j = \min\{b_k \mid b_i < a_k\} \\ nil, & otherwise. \end{array}\right]$$

That is $next\,(I_i)$ is the interval that ends farthest to the left among all the intervals beginning after the end of I_i (see Fig 1).

To compute $next(I_i)$, $(1 \le i \le n)$, we use the same algorithm used for *maxright* (I_i) (Algorithm 1) with a new step 4. Algorithm compute the function next.

It is easy to see that the given procedure implements the definition of $next\,(I_i)$ with the same complexity as for computing *maxright* (I_i) which is $O(n)$.

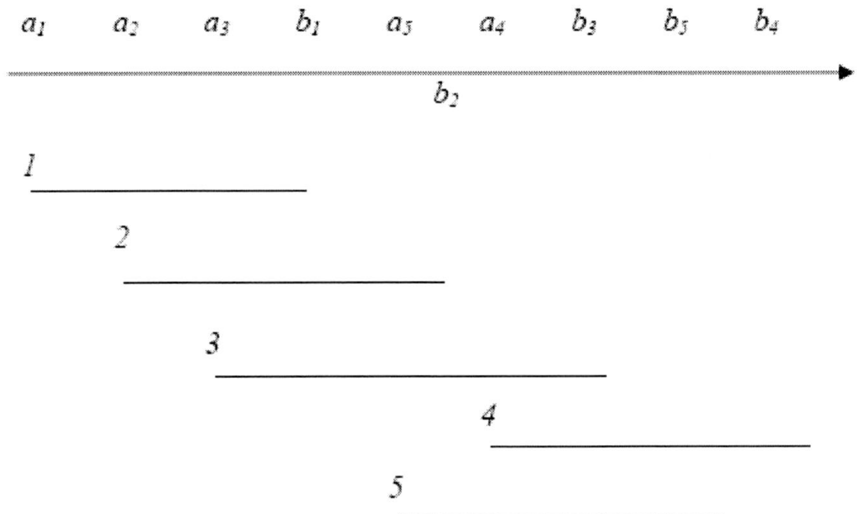

maxright (1) = 3, maxright (2) = 5, maxright (3) = maxright (5) = 4, maxright (4) = nil
maxright (1) = 2, minright (2) = 3, minright (3) = 5 , minright (4) = nil
minright (5) = 4
first({1, 2, 3, 4, 5})=1
next (1) = 5, next (2) = 4, next (3) = next (4) = next (5) = nil

Figure 1. An Example of the *maxright, minright, first* and *next* functions.

5. Further Extensions

This section sketches how the shortest paths algorithm of the previous sections can be used to solve problems where intervals can have zero weight, and how it can be used to solve the version of the problem where we have circular -arcs rather than intervals on a line.

5.1. Zero- Weight Intervals

The astute reader will have observed that the definitions and the shortest paths algorithm of the previous sections can be modified to handle zero -weight intervals as well. However, doing so would unnecessarily clutter the exposition. Instead, we show in what follows that the shortest -paths problem in which some intervals have zero weight can be reduced in linear time to one in which some intervals have zero weight can be reduced in linear time to one in which all the weights are positive. Not only does this simplify the exposition, but the reduction used of independent interest.

Let p_1 be the version of the problem that has zero - weight intervals of S. First observe that in order to solve p_1, it suffices to solve problem p_2 obtained from p_1 by replacing every connected component CC of Z by a new zero weight interval that is the union of the zero - weight intervals in CC (because the label of $I \in z$ in p_1 is the same as the label of $J= \cup_{I \in CC} I$

in p_2). Hence it suffices to show how to solve p_2. In what follows assume that we have already created, in $O(n)$ time, p_2 from p_1.

We next show how to obtain, from p_2, a problem p_3 such that :

i. Every interval in p_3 has a positive weight (and therefore p_3 can be solved by the algorithm of the previous sections .)
ii. The solution to p_3 can be used to obtain a solution to p_2 .

Recall that, by the definition of p_2, the zero - weight intervals in it can not overlap . p_3 is obtained from p_2 by doing the following for each zero weight interval $J=[a,b]$: "cut-out" the portion of the problem between a and b, that is, first erase, for every interval I of p_2, the portion of I between a and b then " pull " a and b together so they coincide in p_3 . This means that, in p_3, J has disappeared , and so has every interval J′ that was contained in J . An interval J'' in p_2 that contained J, or that properly overlapped with J, gets shrunk by the disappearance of its portion that used to overlap with J. For example, if we imagine that the situation in fig 1 describes problem p_2, and that J is (sag) interval I_4 in Figure 1 (so I_4 has zero weight) , then "cutting " I_4 results in the disappearance of I_2 and I_3 and the " bringing together " of I_1 and I_{10} so that, in the new situation, the right endpoint of I_1 coincides with the left endpoints of I_{10} .

5.2. Implementation Note

The above described cutting - out process of the zero weight intervals can be implemented in linear time by using a linked list to do the cutting and pasting . In particular, if in p_2 an interval I of positive weight contains many zero-weight intervals J_1 , . . . , J_k , the cutting - out of these zero -weight intervals does not affect the representation we use for I (although in a geometric sense I is " shorter " afterward, as far as the linked list representation is concerned, it is unchanged). This is an important point, since it implies that only the end points contained in a J_k are affected by the cutting out of that J_k , and such an end point gets updated only once because it is not contained in any other zero - weight interval of p_2 (recall that the zero weight intervals of p_2 are pair wise non overlapping) .

By definition, p_3 has no zero - weight intervals . So suppose p_3 has been solved by using the algorithm we gave in the earlier sections . The solutions to p_3 yields a solution to p_2 in the following way .

- If an interval I is in p3 (i.e., I has not been cut-out when p3 was obtained from p2) , then it label in p2 is exactly the same as its label in p3 .
- Let J = [a,b] be a zero weight interval which was cut out from p2 when p3 was created. (In p3 , a and b coincide, so in what follows when we refer to " a in p3 " we are also referring to b in p3.) . For each such J= [a,b] compute in p3 the smallest label of any interval of p3 that contains a : this is the label of J in p2. This computation can be done for all such J's by one linear - time scan of the endpoints of the active intervals for p3 .
- Suppose I is a positive weight interval of p2 that was cut- out when p3 was created , because it was contained in a zero - weight interval J of p2 that was cut out when p3

was created , because it was contained in a zero -weight interval J of weight interval J of p2. Then the label of I in p2 is equal to (weight of I) + label of J in p2) .

5.3. Circular -Arcs

The version of the shortest - paths problem where we have circular - arcs on a circle *C* instead of intervals on a straight line can be solved by two applications of the shortest paths algorithm for intervals : Suppose $I_1 = [a,b]$ is the source circular - arc, where *a* and *b* are now positions on circle *C*. (We use the convention of writing a circular - arc as a pair of positions on the circle such that when going from the first position to the last travel in the clock wise direction .

It is not hard to see that the following linear shortest path problem on circular - arc graphs.

- Create a problem on a shortest line of n interval problem . Intervals that contain a are not included twice in the straight line problem : only there first appearance on the clock wise trip is used, and they are " truncated " at a (so that on the line, they appear to begin at a , just like the source I1). Then solve the straight line problem so created , by using the algorithm for the interval case. The computation of this step gives each circular - arc a label.
- Repeat the above step with a playing the role of b, and "counter clock wise" playing the role of " clock wise " .
- The correct label for a circular - arc is the smaller of the two labels, computed above , for the intervals corresponding to that arc.

6. Conclusions

We have given a linear processor CREW PRAM algorithm for determining the shortest paths in an interval graphs which runs in $O(n)$. Our motivation for considering graphs was to see if they can be used to solve the shortest - path problem for interval graphs. Our algorithm solves this problem optimally in $O(n)$ time, where n is the number of intervals in a graph. The $n \log n$ term time complexity is obtained from a preprocessing sorting step when the sorted list of endpoints is not given as a part of the input.

Acknowledgements

The author is highly thankful to anonymous referees for helpful comments for improving the paper. The author is also thankful to INSA- New Delhi, Govt. of India for providing the financial support for the project in the scheme of INSA Visiting Scientist Fellowship.

REFERENCES

Aho, A. V., Hopcroft, J. E. and Ullman 1974. The Design and Analysis of Computer Algorithms, *Addison-Wesley, Reading, MA.*

Atallah, M. J. and Chen D.Z. 1989. An optimal parallel algorithm for the minimum circle cover problem. *Information Processing Letters*,32:159-165.

Bertossi, A. A. 1988.Parallel Circle cover algorithms. *Information Processing Letters*, 27:133-139.

Booth, K. S. and Luekher, G. S. 1976. Testing for the consecutive onesproperly, interval graphs ,and graph planarity using PQ-tree algorithms. *Journal of Computer and System Sciences*,13:335-379.

Chen, D. Z. and Lee, D. T. 1994.Solving the all pair shortest path problem on interval and circular-arc graphs. *Proceedings of the 8th International Parallel Processing symposium,* Cancun, Mexico:224-228.

Gabow, H. N. and Tarjan, R. E.1985. A linear time algorithm for a special case of disjoint set union . *Journal of Computer and System Sciences*, 30:209-221.

Golumbic, M. C. 1980. Algorithmic Graph Theory and Perfect Graphs. *Academic Press*, New York.

Gupta, U. I., Lee, D. T. and Leung J. Y-T. 1982. Efficient algorithms for interval graphs and circular arc graphs. *Networks*, 12:459-467.

Ibarrra, O. H. Wang , H. Zheng, Q. 1992. Minimum cover and single source shortest path problems for weighted interval graphs and circular arc graphs . Proceeding of the 30th Annual Allerton Conference on Communication , Control and Computing, University of Illinois, Urbana:575-584.

Lee, C. C. and Lee, D. T. 1984. On a circle cover minimization problem. *Information Processing Letters* 18:109-115.

Mishra, P. K. 2004. An Efficient Parallel Algorithm for Shortest Paths in Planar Layered Digraphs. *Journal of* Zhejianh *University, SCIENCE* 5(5):518-527.

Mishra, P. K. and Sharma, C. K. 2002. An Efficient Implementation of Scaling Algorithms for the Shortest Paths Problem, *News Bulletin of Calcutta Mathematical Society Vol. 27 : 342-351.*

Mishra, P. K. and Sharma, C. K. 1997. A Computational Study of the Shortest Path Algorithms in C- Programming Language, *Proc. Fifth International Conference on Applications of High Performance Computing in Engineering*, Santiago de Compostela, Spain.

Mishra, P. K. 2004. Optimal Parallel Algorithm for Shortest Paths Problem in Interval Graphs,. *Journal of* Zhejianh *University, SCIENCE* 5(9):1135-1143.

In: Progress in Chaos Complexity Research
Editors: Franco F. Orsucci and Nicoletta Sala
ISBN: 978-1-60456-375-7

Chapter 4

DETECTING LOW DIMENSIONAL CHAOS IN SMALL NOISY SAMPLE SETS

Nicolas Wesner *
Mazars Actuariat, Economix, Paris, France

ABSTRACT

A new method for detecting low dimensional chaos in noisy small sample sets is presented. A quantity that can be interpreted as a measure of the degree of determinism or nonlinear mean predictability in a time series is defined on the basis of the embedding theorem and the method of time delays (Takens 1981). Numerical experiments on stochastic and chaotic processes show that, in application to very short time series, this method is effective, while traditional approaches such as the false nearest neighbors method have difficulties.

INTRODUCTION

The concept of deterministic chaos implies the generic possibility that an apparently random phenomenon is actually generated by a deterministic process and thus concerns all domains of scientific research. It has significantly changed the approach of data analysis in providing new tools for distinguishing random time series from deterministic ones. For this purpose, time series are analyzed not only in the time domain but in the phase space too. This geometrical approach relies on the reconstruction of data in the phase space, and Takens' theorem (Taken 1981) provides theoretical grounds for it. The first tools such as the correlation integral (Grassberger and Procaccia 1983), which measures spatial correlations in the phase space, or the Lyapunov exponent (Wolf et al. 1985) which measures sensitive dependence on initial conditions, were designed for large data sets (over 10000 observations) and perform poorly with small data sets.

* Nicolas Wesner: wesner@mazars.fr

Since, researchers have come up with numerous methods for observing or measuring nonlinear determinism in relatively shorter time series. Most of them are based on the observed continuity of trajectories in the state space (Aleksic 1990, Kennel et al. 1992, Wayland et al. 1993, Cao 1997, Zbilut et al. 1998). Like the correlation dimension, they permit us to determine the minimum embedding dimension of a time series, that is the minimal dimension of the state space for which the intensity of space correlations is the highest. The basic idea behind this approach is that for a chaotic process, this minimal dimension is the integer immediately above the fractal dimension of the attractor, whereas for random numbers, no such dimension exists because the intensity of spatial correlations is low and does not depend on the embedding dimension. Those methods are more or less complex but seem to work well. Nevertheless, all those methods are designed for relatively large data sets and far nothing is known about their efficacy for very short samples in the presence of observational noise.

The aim of this paper is to present a very simple method for detecting low dimensional deterministic structure in small sample sets infected by additive noise. The need for such quantitative techniques is particularly important in social sciences like economics where data are noisy by nature and often observed on low frequency (quarterly, yearly) so that few observations are available. The method presented here relies on the Takens embedding theorem and on the property of phase space continuity which is characteristic of well reconstructed deterministic time series. The method is tested for short time series generated by various stochastic processes and noisy chaotic processes. Results show that it overcomes the shortcomings of traditional techniques, and is able to clearly discriminate long memory and non normal stochastic processes from chaotic time series contaminated by noise. Finally the method is applied to real world data.

2. Methodology

2.1. Reconstruction of Dynamics by the Method of Time Delays

Let $s(t)$ $(t=1,...,N)$ be a time series that is believed to be generated by an unknown or unobservable deterministic process. Following Brock (1986), $s(t)$ is said to have a smoothly deterministic explanation if there exist a system $\{M,F,h\}$ such as :

$$F : M \to M, \quad x(t+1) = F(x(t)) \tag{1}$$

where F is an unobservable, smooth (i.e. twice differentiable almost everywhere) mapping, M is a d dimensional manifold and $x(t) \in M$ is the state of the system at time t.

The observed time series is related to the dynamical system by the measurement function h :

$$F : M \to M, \quad s(t) = h(x(t)) \tag{2}$$

where h is an unobservable smooth mapping,

State space reconstruction is aimed to recover information about the unknown system *{M,F,h}* only from the observed time series *s(t)*. The most common technique used for this purpose is the method of time delays. The basic idea is that the past and future of a time series contain information about unobserved state variables that can be used to define a state at the present time. Concretely, lagged observations of the original time series are used to construct a new series of vectors *y(t)* called m-histories:

$$y(t)=(s(t),s(t-\tau),...,s(t-\tau(m-1))) \quad (3)$$

where m is the embedding dimension and τ the time delay usually fixed to one.

Takens studied the delay reconstruction map Φ which maps the states of a *d* dimensional dynamical system into m dimensional delay vectors :

$$\Phi : M \rightarrow M' = IR^{m^2} \quad (4)$$
$$\Phi(x(t)) = [\,H(x), H(F^{\tau}(x)), ..., H(F^{\tau(m-1)}(x))\,]$$

Takens demonstrated that with $m \geq 2d+1$, Φ is generically an embedding, that is a diffeomorphic mapping between a compact set in a finite or infinite dimensional space and a subspace of finite dimension (see Sauer et al. 1992 for a more formal definition of an embedding).

The main point is that if Φ is an embedding then a smooth dynamics *G*, equivalent to the original *F*, is induced on the space of the reconstructed vectors :

$$G(y(t)) = \phi^{\circ}F^{\circ}\phi^{-1}(y(t)) \quad (5)$$

G is a diffeomorfism to *F* and is called topologically conjugate to *F*. That is *G* conserves the same properties as *F*. Therefore, with τ=1:

$$G : IR^{m^2} \rightarrow IR^{m^2}, \quad (6)$$
$$y(t+1) = G(y(t))$$

The reconstructed states can be used to estimate *G*, and since *G* is equivalent to the original dynamics F, those can be used to extract information about the underlying, unknown system.

Thus, according to the theorem of Takens, if the time series *s(t)* has a deterministic explanation, then for any pair of points *(y(i),y(j))*, for a small α>0, and for an adequate choice of m, there exists a small δ>0 so that:

$$\text{if } \| y_{(i)} - y_{(j)} \| < \alpha, \qquad \text{then, } \| G(y_{(i)}) - G(y_{(j)}) \| < \delta \quad (7)$$

This property specific to deterministic time series is actually a property of a continuous mapping : images of close points are close. This property of "phase space continuity" can be used to distinguish between chaotic and stochastic time series or can be exploited to make nonlinear mean forecasts (Farmer and Sidorowich 1987).

It is important to notice that the condition that the measurement function h be smooth entails that the result of the theorem does not necessarily hold even approximately when the data are contaminated by noise. For additive and dynamical noise, the Takens embedding theorem fails and one can no longer assume that there exists a deterministic map characterizing the time evolution of *s(t)*. However, numerical investigations of toy models indicate that reconstruction even from heavily noise contaminated series can be performed quite successfully.

2.2. A Measure of Determinism

The measure of determinism presented here relies on the reconstruction scheme described above. The basic idea is to measure the extent to which the time series considered verify the property of "phase space continuity", by observing only the dynamics of nearest neighbors. This quantity can also be interpreted as a measure of nonlinear mean predictability since phase space continuity can be exploited by local predictor (Farmer and Sidorowich 1987). For instance, it is assumed that nearest neighbors whose images are nearest neighbors satisfy the continuity property (7). That is, y(i) and y(j) satisfy the continuity property (7) if:

$$y(j) = \text{argmin} \{ \| y_{(i)} - y_{(s)} \|, s \neq i, = m, \ldots, M\}, \quad (8)$$

and if :

$$y(j+1) = argmin \{ \| y_{(i+1)} - y_{(s)} \|, s \neq i+1, = m, \ldots, M\}, \quad (9)$$

Indeed for those points :

$$\| y_{(i)} - y_{(j)} \| = r(i) \text{ and } \| y_{(i+1)} - y_{(j+1)} \| = r(i+1), \quad (10)$$

where r(s) is the minimum distance between y(s) and another vector in the phase space.

Thus, if *y(i)* and *y(j)* satisfy (8) and (9), then it is possible to choose arbitrary small α and δ for which *y(i)* and *y(j)* verify (7).

Numerical experiments show that, for short time series, the proportion of points satisfying those properties grows with the embedding dimension. So the measure of determinism D proposed here is defined as follows:

$$D = [\text{number of pairs of points } y(i) \text{ and } y(j) \text{ satisfying (8) and (9)}] / (N-m+1)m \quad (11)$$

where N is the number of observations and m is the embedding dimension.

Except for special cases like the tent map, doubling map or logistic map where $D=1/(2m)$ for large N, it is difficult to derive theoretical values of *D*. Nevertheless, for independent and stationary data, the probability that nearest neighbors in the reconstructed state remain nearest neighbors under time evolution decreases when N grows, so *D* is expected to be close to 0 for *N* sufficiently large.

3. Numerical Experiments

3.1. Applying Traditional Methods to Small Sample Set

The false nearest neighbors approach (Kennel et al. 1992) is perhaps the most popular method for determining the minimum embedding dimension of a time series. False neighbors are defined as points apparently lying close together due to projection that are separated in higher embedding dimensions.

Nearest neighbors $y(i)$ and $y(j)$ are declared false if:

$$| s(i+1) - s(j+1)| \,/\, \| y_{(i)} - y_{(j)} \| > R_{tol} \tag{12}$$

or if :

$$[\| y_{(i)} - y_{(j)} \|^2 - | s(i+1) - s(j+1)|^2] \,/\, R^2_A > A^2_{tol} \tag{13}$$

where

$$R^2_A = 1/N \, \Sigma_k \, | s(k) - \langle s \rangle |^2, \ \langle s \rangle \text{ is the mean of } s(t). \tag{14}$$

For a deterministic process, the percentage of false nearest neighbors should drop to zero or some acceptable small numbers by increasing the embedding dimension. In the following application as in most studies, R_{tol} is set to 10 and A^2_{tol} to 2.

Percentages of false nearest neighbors were calculated for time series of 200 observations from the Standard Cauchy distribution, Fractionally Integrated moving average process, chaotic Mackey Glass system and Mackey Glass process with additive Cauchy noise.

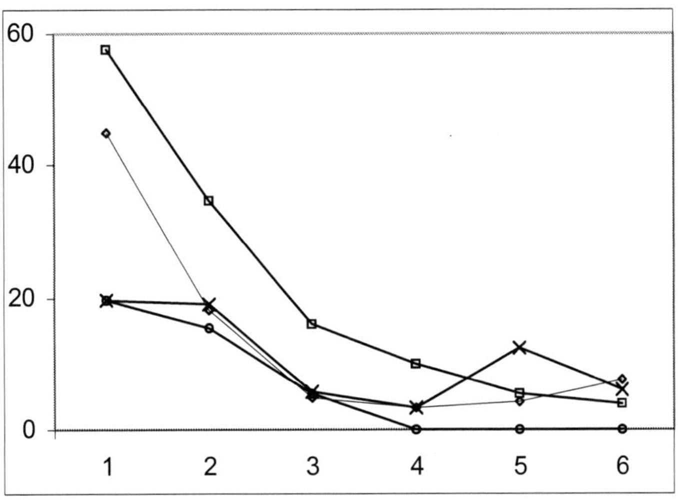

Figure 1. Percentage of false neighbors as function of the embedding dimension m. Squares and diamonds for Cauchy and fractional noise, rounds for the chaotic process and crosses for the noisy chaotic time series.

The Standard Cauchy distribution is the distribution of the quotient of independent standard normal random variables. The tails of the distribution, although exponentially decreasing, are asymptotically much larger than any corresponding normal distribution. Its mode and median are zero, but the expectation value, variance and higher moments are undefined:

$$f(s)= [\pi(1+s^2)]^{-1} \quad (15)$$

Fractional Integrated noise displays the property of long memory which means that the autocorrelations decay so slowly that their sum does not converge. More generally, long memory can be seen as a form of long range dependence. The following process is an example of fractional noise known as an ARFIMA(0,0.6,1) process:

$$(1-B)^{0.6}s(t) =(1-aB)\varepsilon(t) \quad (16)$$

where $\varepsilon(t)$ *is a white noise process, and B is the backshift operator such that* $B^j s(t) = s(t-j)$.

The chaotic Mackey Glass system, is defined as follows:

$$ds/dt = 0.2s(t-\tau)/[1+s(t-\tau)^{10}] - 0.1s(t),\ \tau=30 \quad (17)$$

The noisy chaotic time series was constructed in adding a Standard Cauchy white noise to the Mackey glass process. The influence of the additive noise was weighted by a coefficient which represents a fraction of the standard deviation of x(t).

$$s(t)=x(t)+0.01\sigma\varepsilon(t) \quad (18)$$

where ε (t) is a white noise with a standard Cauchy distribution, x(t) is generated from a Mackey Glass process, and σ is the standard deviation of x(t).

As can be seen on Graphic 1, the false nearest neighbors approach is unable to discriminate between fractional noise, Cauchy noise and noisy chaotic time series. Indeed, the percentages of false neighbors decrease with m and are relatively low (under 5%) for the stochastic processes and those are not lower for noisy chaotic data.

The shortcomings of traditional methods of dimension estimation have already been noticed in the literature for autocorrelated noise and for small sample set. Indeed, it is well known that temporal correlations in time series lead to spuriously low dimension estimates (Theiler 1986, Abraham et al. 1986, Havstadt and Ehlers 1989, and Osborne and Provenzale 1989). Moreover, the sample size needed grows exponentially fast as the dimension increases (Mayer Kress 1987, Smith 1988, Kantz and Schreiber 1995), thus the lack of a sufficient number of observations has the same effect as the presence of temporal correlations: it produces spuriously low dimension estimates and prevents from discriminating correctly random numbers from deterministic chaos. Finite sample problem as well as the presence of autocorrelations can be treated, notably in using the method of surrogate data (Theiler et al. 1992). Nevertheless, the negative results obtained here seem to indicate that even surrogate data analysis should not help for discriminating between chaotic time series with additive non normal noise and complex stochastic processes in using traditional methods.

3.2. A Method Adapted to Small Data Set

3.2.1. Convergence to Theoretical Value

The logistic map is a good example of how complex, chaotic behavior can arise from very simple non-linear dynamical equations.

$$s(t+1)=as(t)(1-s(t)), \qquad (19)$$

where a=4 and $s(0)\in [0,1]$.

Like the tent map, this application has a symmetric invariant density on the interval [0,1] and two equally probable pre-images for each state. For the chaotic state defined above, the probability measure that gives the long-run proportion of time spent by the system in the various regions of the attractor, and which corresponds to the beta distribution for chaotic the state, permits us to derive the asymptotic value of *D*, that is: $D=1/2m$.

In order to investigate the speed of convergence to its asymptotic value, the quantity was calculated for number of observations *N* growing from 1 to 30 and for embedding dimension from 1 to 4. As can be seen on graphic 2 the quantity quickly converges around the theoretical value. The results shows also that *D* depends in a cyclical way on *N*:

$$D= (int(N/2)/N)\,/\,2m, \qquad (20)$$

where int(x) is the higher or lower integer closest to x. Whether the integer is the lower or the higher depends on the embedding dimension.

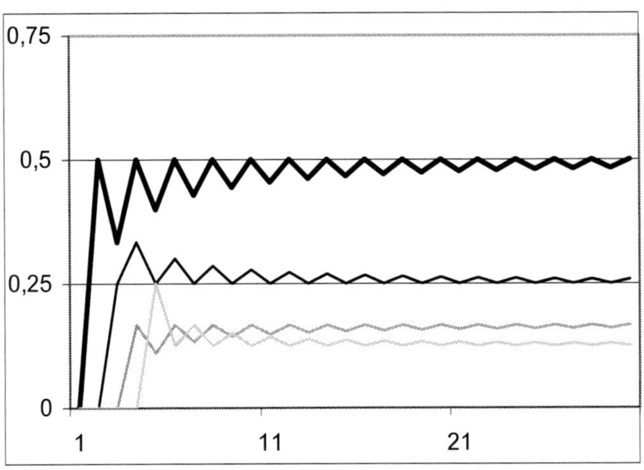

Figure 2. The quantity D for number of observations N growing from 1 to 30. Large black line for m=1, thin black for m=2, grey for m=3 and white for m=4.

This cyclical behavior is due to the map's stretching-and-folding of the space on which it is defined. As can be seen on table 1, the estimation errors become very low over 200 observations. So those results seem to indicate that the measure could be applied to small sample sets. However it is important to notice that the behavior of the quantity D as well as the scaling law of the estimation errors observed here are specific to the logistic map and do not represent a general result. The next sections explore the ability of the quantity to distinguish between complex stochastic processes and noisy chaotic time series with small sample sets.

Table 1. Differences between theoretical and estimated values of *D* for different values of the embedding dimension and different sizes of the sample set

N	m=1	m=2	m=3	m=4	m=5	m=6
11	0.0454	0.0227	0.0151	0.0114	-0.0091	0.0076
51	-0.0098	0.0049	-0.0033	0.0024	-0.002	0.0016
101	-0.0049	0.0025	-0.0016	0.0012	-0.001	0.0008
201	-0.0025	0.0012	-0.0083	0.0006	-0.0005	0.0004
501	-0.001	0.0005	-0.0003	0.0002	0.0002	0.0002
999	-0.0005	0.0002	-0.0002	0.0001	-0.0001	-0.0001

3.2.2. Stochastic Processes

As in the previous application, the quantity *D* was calculated for Standard Cauchy noise and Fractional noise for time series of 200 observations. The results were compared to those for normal white noise and colored noise (with an autocorrelation coefficient of 0.9). The higher values of *D* obtained for fractional noise and Cauchy noise may be due respectively to long term correlations and to the presence of extreme values (outliers). Nevertheless, in all cases, the quantity D does not exceed 0.1 (see table 2).

Table 2. The quantity D calculated for time series of 200 observations generated by various stochastic processes, for embedding dimensions from 1 to 6

	1	2	3	4	5	6
Gaussian white noise	0.005	0.03	0.047	0048	0.056	0.049
Gaussian colored noise	0	0.028	0.035	0.055	0.043	0.049
Fractional noise	0.01	0.025	0.067	0.071	0.072	0.061
Standard Cauchy noise	0	0.038	0.071	0.071	0.069	0.063

3.2.3. Chaotic Processes Infected with Additive Complex Noise

The quantity was calculated for time series generated by the Hénon map, Lorenz system and Mackey Glass processes with different levels of additive Cauchy and Fractional Noise.

The Hénon Map is defined as follows: $s(t+1) = 1 - as(t) + s1(t)$, $s1(t) = bs(t)$, $a=1.4$ *et* $b=0.3$

The results shown in table 3 indicate that, for strictly deterministic processes, the quantity *D* becomes superior to 0.1 for a sufficient high value of m. Moreover, for a small amount of additive Cauchy or Fractional noise (1 or 2%), the method is still able to recognize the deterministic structure of the Mackey Glass system.

In summary, the results show that, for data sets of 200 observations, while traditional methods such as the false nearest neighbors approach have difficulties, the method presented here can clearly distinguish between various forms of stochastic processes and time series generated by chaotic systems even contaminated by complex additive noise.

Table 3. The quantity D for various chaotic and noisy chaotic time series

	1	2	3	4	5	6
Hénon	0.115	0.276	0.215	0.213	0.134	0.115
Mackey Glass	0.025	0.055	0.069	0.084	0.133	0.124
M-G +1% Cauchy Noise	0.015	0.058	0.062	0.072	0.125	0.124
M -G +2% Fractional Noise	0.03	0.05	0.066	0.085	0.128	0.12
M -G +2% Cauchy Noise	0.01	0.055	0.057	0.08	0.108	0.114
M -G +5% Cauchy Noise	0	0.04	0.066	0.07	0.077	0.089

3.2.4. Real World Data

Santa Fe Institute Prediction Competition

The method was applied to a time series from the Santa Fe Institute Prediction Competition: 200 observations of the fluctuations in a far infrared laser from a laboratory experiment.

The quantity D is superior to the threshold value 0.1 for m superior to 1 (see table 4). Thus, according to those results, it appears that the time series contains nonlinear dependences that could be exploited by local predictor.

Stock Market Prices: Dow Jones Yearly Returns

The presence of nonlinear determinism in asset prices dynamics is of great importance in finance since it indicates a certain degree of predictability, and thus is susceptible to invalidate the informational efficiency hypothesis (Fama 1965). Techniques of chaos detection as well as local predictors have largely been applied to economic and financial time series but all those works have only studied relatively long time series, that is weekly or daily observations.

Table 4. The measure of determinism D for real world data

	1	2	3	4	5	6
SFI competition data	0	0.163	0.151	0.142	0.142	0.126
Dow Jones yearly returns	0.005	0.022	0.051	0.038	0.031	0.028

The quantity D was calculated for the yearly returns of the Dow Jones index. The analysis is multidimensional in the sense that independent variables are used in addition to lagged variables. From 1896 to 2003, yearly returns are calculated for the higher and the lower prices of the year and for the closing price of the last day. Thus, embedding vectors y(t) are defined as follows:

$$y(t)=(r\{l\}(t),...,r\{l\}(t-m+1)),r\{h\}(t),...,r\{h\}(t-m+1)),r\{c\}(t),...,r\{c\}(t-m+1)), \quad (21)$$

where r{l}(t), r{h}(t), r{c}(t) denote the returns calculated for lower, higher and closing yearly prices of the index.}

Results indicate the absence of nonlinear deterministic dependences exploitable for conditional mean forecast and are thus in line with the informational efficiency hypothesis.

Conclusion

A method for discriminating between stochastic processes and deterministic processes infected by noise was developed. Like other approaches cited in the introduction, this method does not provide an absolute statistical test for determinism (if such a test existed), it does provide a quantitative measure of the appropriateness of a deterministic explanation for an observed dynamics. However it has the main advantage that it works well for very short time series and for chaotic processes infected with additive non normal noise and long memory processes. Moreover, this method is very simple, requires few computer resources, and does not contain subjective parameters. As it has be shown, it should be applied to realistic time series and particularly to historical data on low frequency or over short periods of time, for which relatively few observations are available.

References

Abraham N.B., Albano A.M, Das B., De Guzman G., Yong S., Gioggia R.S., Puccioni G.P. and Treddice J.R. (1986) "Calculating the dimension of attractors from small data sets", *Physics Letters*, 114 A, 217-221.

Aleksic Z. (1991) "Estimating the embedding dimension", *Physica D* 52 362-368.

Brock W.A. (1986) "Distinguishing Random and Deterministic Systems: Abriged Version", *Journal of Economic Theory* 40, 168-195.

Cao L. (1997) "Practical method for determining the minimum embedding dimension of a scalar time series", *Physica D* 110, 43-50.

Eckmann J.P., Oliffson S.Kamphorst and Ruelle D. (1987) "Recurrence plots of dynamical system", *Europhysics Letters* Vol. 4 No. 9, 973-977.

Fama E. (1965) "Random walks in stock market prices", *Financial Analysts Journal*, 21(5), 34-109.

Farmer J.D. and Sidorowich J.J. (1987) "Predicting chaotic time series", *Physical Review Letters*,vol.59 8, 845-848.

Grassberger P. and Procaccia I. (1983) "Chararcterization of strange attractors", *Physical Review Letters* 50, 189-208.

Havstad J.W. and Ehlers C.L. (1989) "Attractor dimension of nonstationary dynamical systems from small data sets", *Physical Review* A, 39, 845-853.

Kantz H. and Schreiber T.(1995) "Dimension estimates and physioligical data", *Chaos* 5, 143-154.

Kennel M., Brown R. and Abarbanel H. (1992) "Determining embedding dimension for phase-space reconstruction using a geometrical *construction"*, *Physical Review* A 45, 3403.

Mayer-Kress G. (1987) "Directions in Chaos", In Hao Bai-Lin (Ed) *World Scientific series on Directions in condensed matter Physics*, World Scientific, Singapore.

Osborne A.R. and Provenzale A (1989) "Finite correlation dimension for stochastic systems with power-laws spectra", *Physica D* 35, 357-381.

Sauer T., Yorke J.A., Casdagli M. (1992) "Embodology", *Journal of Statistical Physics* 65(3/4), 193-215.

Smith L.A. (1988) "Intrinsic limits of dimension calculations", *Physics Letters* A, 133, 283-288.

Takens F. (1981) "Detecting strange attractors in fluid turbulence", In Rand D.A., Young L.S. (Eds.) *Dynamical System and Turbulence*, Lecture notes in mathematics, Springer Verlag, Berlin 366-381.

Theiler J. (1986) "Spurious dimension from correlation algorithm applied to limited time series data", *Physical Review* A 34 2427-2432.

Theiler J. Galdrikian B., Longtin A., Eubank S., Farmer J.D. (1992) "Using Surrogate Data to Detect Nonlinearity in Time Series", *Nonlinear Modelling and Forecasting*, SFI Studies in the Sciences of Complexity, Eds. Casdagli Eubank, Vol.XII., 163-188.

Wayland R., Bromley D., Pickett D. and Passamante A. (1993) "Recognizing determinism in a time series", *Physical Review Letters* vol.70 n°5, 580-587.

Wolf A., Swift J.B., Swinney H.L. and Vastano J.A. (1985) "Determining Lyapunov exponents from a time series", *Physica D* {16} 285-317.

Zbilut J.P., Giuliani A. and Webber C.L. (1998) "Recurrence quantification analysis and principal components in the detection of short complex signal", *Physics Letters* A {237} 131.

In: Progress in Chaos Complexity Research
Editors: Franco F. Orsucci and Nicoletta Sala
ISBN: 978-1-60456-375-7

Chapter 5

A SENSITIVITY STUDY ON THE HYDRODYNAMICS OF THE VERBANO LAKE BY MEANS OF A CFD TOOL: THE 3D EFFECTS OF AFFLUENTS, EFFLUENT AND WIND

Walter Ambrosetti[1], Nicoletta Sala[2] and Leonardo Castellano[3]
[1] CNR – Istituto per lo Studio degli Ecosistemi,
Verbania (Italy), Email: w.ambrosetti@ise.cnr.it
[2] Università della Svizzera Italiana, Mendrisio (Switzerland)
Email: nsala@arch.unisi.ch
[3] Matec Modelli Matematici, Milano (Italy),
Email: matec_srl@iol.it ,

ABSTRACT

This short report deals the use of three dimensional CFD (Computational Fluid Dynamics) simulations in order to better understand the complex interactions between the hydrodynamics of a given water body and the chemical, physical, biological and meteorological phenomena.

As the difficulties increase for the availability of the drinkable water and for the life of the aquatic ecosystems, the necessity for limnologists to enlarge even more the extended compass of the instruments and methods they use in their studies becomes essential, [1].

The formulation of multidimensional mathematical models and computer codes for fluid-dynamics and heat and mass transfer in seas, lakes and rivers is not a recent activity [e.g. 2-5]. It started many decades ago; received a great impulse with the publication of the famous report "The Limits to Growth" (1972); and reached the maximum point of creativity at about the second half of the 80th years of past century [e.g. 6-8]. The successive improvements, from that time up to the current days, are mainly due to the tremendous increase of the computational power of the computers that has allowed the scientists to use several hundred or million computational cells but also to implement numerical manipulation of sophisticated models of turbulence (e.g. Large Eddy Simulation; LES), realistic interactions with meteorological events and more complex chemical and biochemical schemes [e.g. 9-11].

The above progress, although very noticeable from the theoretical point of view, are no more sufficient to build up an "universal" model but enough attractive to cultivate the illusion that the reconstruction of real scenarios can be just a matter of computational resources and time.

That should be a fatal and expensive error in that they are two possibilities: a) the exploitation of the (very large) output of a 3D simulation only to extract gross parameters; b) otherwise, the use of this output to extract details at the scale of local points where in situ measurements are available. In the first case, except for trivial errors, the answer can be only a confirm of the goodness of the traditional instruments of analysis. In the second case, the questions arise: how many experimental points are need to collect in order to provide a representative image of the system of interest ? how many experimental and theoretical points are to be found in agreement in order to conclude that the numerical simulation is satisfactory?

The authors of the present paper are convinced that in the field of the limnological studies the better use of sophisticated mathematical models is to open a "dialog" with the water body; i.e. to proceed by analyzing step-by-step the answer of the system at different hydrodynamic, thermal and meteorological loads and using the results of each single simulation to try to forecast the output of the next more complicated one and interpret the discrepancies. Only at conclusion of a careful study of this kind a some overall simulation can give really useful results.

A first attempt to use a 3D-CFD model is now in progress on Lake Maggiore (Italy) based on the availability of 50-years' hydrometeorological, physical, chemical and biological daily records. That application is inspired at the above philosophy.

REFERENCES

Ambrosetti W., Barbanti L., Sala N. (2003), "Residence Time and Physical Processes in Lakes", *J. Limnol.*, 62 (Suppl. 1), 2003, pp 1-15.

Orlob G.T. (1967), *Prediction of Thermal Energy Distribution in Streams and Reservoirs*, Water Resources Engineers, Inc., Walnut Creek, California, Technical Report prepared for California State Department of Fish and Game, June 30, 1967.

Cheng R.T., Tung C. (1970), "Wind-Driven Lake Circulation by Finite Element Method", *Proceedings of the 13th Conference on Great Lakes Research*, Buffalo, New York, pp. 891-903.

Dailey J.E., Harleman D.R.F. (1972), *Numerical Model for the Prediction of Transient Water Quality in Estuary Networks*, MIT Department of Civil Engineering, Report No. MITSG 72-15, Cambridge, Mass.

Castellano L., Dinelli G. (1975), "Experimental and Analytical Evaluation of Thermal Alteration in the Mediterranean", *Int. Conference on Mathematical Models for*

Environmental Problems; University of Southampton, September 8-12, 1975, Pentech Press (London).

Blumberg A.F., Mellor G.L. (1987), "A Description of a Three-Dimensional Coastal Ocean Circulation Model", in *Three-Dimensional Coastal Ocean Models*, Heaps N. (ed.), Amer. Geophys. Union, pp. 1–16, 1987.

Hunter, J. R. (1987). "The Application of Lagrangian Particle-Tracking Techniques to Modelling of Dispersion in the Sea", in *Numerical Modelling: Applications to Marine Systems*, Noye J. (ed.), Elsevier Science Publishers, B.V., North-Holland.

Leendertse, J. J. (1989). "A New Approach to Three-Dimensional Free-Surface Flow Modeling", *R-3712-NETH/RC*, Rand Corporation, Santa Monica.

Huisman J., Jöhnk K.D., Sommeijer B. (2003), *Simulation of the Population Development of the Toxic Cyanobacterium Microcystis in Lake Nieuwe Meer under Proposed Heated Water Inflow Scenarios*, Report for IVL Svenska Miljöinstitutet AB and NUON.

Yue W., Lin C.-L., Patel V.C. (2004), "Large Eddy Simulation of Turbulent Open-Channel Flow with Free Surface Smulated by Level Set Method", *Physics of Fluids*, 17(2), 2005, pp. 1-12.

Wang P., Song Y.T., Chao Y., Zhang H. (2005), "Parallel Computation of The Regional Ocean Modeling System", *The International Journal of High Performance Computing Applications*, 19(4), Winter 2005, pp. 375–385.

In: Progress in Chaos Complexity Research
Editors: Franco F. Orsucci and Nicoletta Sala

ISBN: 978-1-60456-375-7

Chapter 6

ALAN TURING MEETS THE SPHINX: SOME OLD AND NEW RIDDLES[1]

Terry Marks-Tarlow*
Private Practice, Santa Monica, California

ABSTRACT

Freud's interpretation of the Oedipus story was the cornerstone of classical psychoanalysis, leading early psychoanalysts to seek repressed wishes among patients to kill their fathers and mate with their mothers. This literal interpretation overlooks a key feature of the Oedipus story, the riddle of the Sphinx. This paper re-examines the Sphinx's riddle – "What walks on four legs in the morning, two legs at noon, and three legs in the evening?" – as a paradox of self-reference. The riddle is paradoxical, by seeming to contradict all known laws of science, and self-referential, in that its solution depends upon Oedipus applying the question to himself as a human being. By threat of death, Oedipus must understand that morning, midday and evening refer not literally to one day, but metaphorically to stages of life. This paper links ancient myth with contemporary computational studies by interpreting the capacity for self-reference as a Universal Turing Machine with full memory, both implicit and explicit, of its own past. A cybernetic perspective dovetails with research on the neurobiology of memory, as well as with cognitive studies derived from developmental psychology. Mental skills required for self-reference and metaphorical thinking signal internal complexity and mature cognition necessary to enter the arena of modern self-reflective consciousness.

Keywords: Paradox of self-reference, complexity, psychoanalysis, second-order cybernetics, Universal Turing Machines.

There is an ancient folk belief that a wise magus can be born only from incest; our immediate interpretation of this, in terms of Oedipus the riddle solver and suitor of his

---

[1] An earlier version of this paper appeared under the title, "Riddle of the Sphinx Revisited," in the Electronic Conference of the Foundations of Information Sciences, May 6-10, 2002

* Terry Marks-Tarlow: (310) 458-3418; markstarlow@hotmail.com

own mother, is that for clairvoyant and magical powers to have broken the spell of the present and the future, the rigid law of individuation and the true magic nature itself, the cause must have been monstrous crime against nature – incest in this case, for how could nature be forced to offer up her secrets if not by being triumphantly resisted – by unnatural acts?

--From Frederick Nietzsche's, The Birth of Tragedy

INTRODUCTION

Throughout history, mythology has inspired the psychology of everyday life at an implicit level. That is, myths help to organize cultural categories and mores by providing roles, models and stories about life in the past, plus rules for future conduct. Whereas ancient and traditional peoples may have experienced myths quite literally (e.g., Jaynes, 1976), the development of the social sciences has shifted steadily to a more symbolic and self-referential focus. Especially since Jung (e.g., 1961), our narratives often examine how myths illuminate the inner world and culture of the mythmakers themselves.

If one myth rises above all others to signal entry into modern consciousness, it is that of Oedipus. This tale has been analyzed throughout the millennia by well-known thinkers as Aristotle, Socrates, Nietzsche, Lévi-Strauss, Lacan and Ricoeur. Some (e.g., Lévi-Strauss, 1977; Ricoeur, 1970) have understood the myth as the individual quest for personal origins or identity; others (e.g., Aristotle, 1982, Nietzsche, 1871/1999) have used sociopolitical and cultural lenses to focus on the tale's prohibitions against the very taboos it illustrates. Prohibitions against infanticide, patricide and incest helped to establish the modern day state, partly by erecting boundaries to protect society's youngest and most vulnerable members, and partly by serving as a kind of social glue to bind individuals into larger collective units. From an evolutionary vantage point, these prohibitions have prevented inbreeding, while maximizing chances for survival and healthy propagation within the collective gene pool.

Perhaps the most prominent analyst of the Oedipus myth has been Sigmund Freud. At the inception of psychoanalysis, this myth proved central to Freud's psychosexual developmental theory as well as his topographical map of the psyche. That this tragic hero killed his father and then married and seduced his mother occupied the psychological lay of the land, so to speak, immortalized as the "Oedipal complex." Whereas Freud (1900) viewed the myth quite literally, in terms of impulses and fantasies towards real people, his successor Jung (1956) interpreted it more symbolically, in terms of intrapsychic aspects of healthy individuation.

The purpose of this paper is to revisit early origins of psychoanalysis that pivot around the Oedipal myth in order to re-examine the narrative from a second-order cybernetic point of view. Cybernetics is the study of information; second-order cybernetics views information science self-referentially by implicating the observer within the observed (see Heims, 1991). From the vantage point of self-reference, the Oedipus story yields important clues about how the modern psyche became more complex via an increased capacity for self-reflection.

In sections to follow, I briefly review the Oedipus myth itself and examine the shift from a literal Freudian interpretation to a more symbolic Jungian one within the early history of psychoanalysis. Then I reason to a new level of abstraction, by applying the approach of Lévi-Strauss to treat the myth structurally. I view the Sphinx's riddle as a paradox of self-reference and argue that both the riddle of the Sphinx and the life course of Oedipus bear structural similarities that signify the self-reflective search for origins. Furthermore, I show how Freud's interest in the Oedipus myth was self-referentially re-enacted in real life through his struggles for authority with Carl Jung.

Next I follow Feder (1974) to examine Oedipus clinically. Oedipus' relentless search for his origins, combined with his ultimate difficulty accepting what he learns, appears at least partly driven by psychobiological symptoms of separation and adoption trauma combined with the physical abuse of attempted murder by his biological father. In the process, I link contemporary research on the psychoneurobiology of implicit versus explicit memory with a cybernetic perspective and the power of Universal Turing Machines able with full access to implicit and explicit memory. Finally, I claim that cognitive skills necessary to move developmentally from concrete to metaphorical thinking, and eventually to full self-actualization, relate to implicit cognition within Lakoff's (1999) embodied philosophy as well as to mature, abstract cognition within Piaget's (e.g., Flavell, 1963) developmental psychology.

While I refer to Sigmund Freud amply throughout this paper, my purpose is primarily historical and contextual. I do not intend to appeal to Freud as the ultimate authority so much as the originator of psychoanalysis and precursor to contemporary thought and practice. Especially since Jeffrey Masson (1984) documented Freud's projection of his own neuroses onto his historical and mythological analyses, including the invention of patients to justify his theories, Freud largely has been de-centered, if not dethroned, within most contemporary psychoanalytic communities. Yet I hope my contemporary reading of Oedipus can help to reinstate the majesty of this myth to the human plight, without sacrificing the many gains and insights gleaned by psychoanalysts and other psychotherapists since Freud's time.

The Myth of Oedipus

The myth of Oedipus dates back to Greek antiquity. King Laius of Thebes was married to Queen Jocasta, but the marriage was barren. Desperate to conceive an heir, King Laius consulted the oracle of Apollo at Delphi, only to receive a shocking prophecy – the couple should remain childless. Any offspring of this union would grow up to murder his father and marry his mother. Laius ordered Jocasta confined within a small palace room and placed under strict prohibitions against sleeping with him.

But Jocasta was not to be stopped. She conceived a plot to intoxicate and mate with her husband. The plot worked, and a son was born. Desperate to prevent fulfillment of the oracle, Laius ordered the boy's ankles be pinned together and that he be left upon a mountain slope to die. But the shepherd earmarked to carry out this order took pity on the boy, delivering him instead to yet another shepherd who took him to King Polybus in the neighboring realm of Corinth. Also suffering from a barren marriage, Polybus promptly adopted the boy as his own. Due to his pierced ankles, the child was called "Oedipus." This name, which translates either to mean "swollen foot" or "know-where," is telling, given Oedipus' life-long limp plus relentless search to "know-where" he came from. I return to the self-referential quality of Oedipus' name in a later section.

As Oedipus matured, he overheard rumors that King Polybus was not his real father. Eager to investigate his true heritage, Oedipus followed in the footsteps of his biological father to visit the oracle at Delphi. The oracle grimly prophesized that Oedipus would murder his father and marry his mother. Horrified, Oedipus attempted to avoid this fate. Still believing Polybus his real father, he decided not to return home, but instead took the road from Delphi towards Thebes, rather than back to Corinth.

Unaware of the underlying truth, Oedipus met his biological father at the narrow crossroads of three paths separating and connecting the cities of Delphi, Corinth and Thebes. King Laius ordered the boy out of the way to let royalty pass. Oedipus responded that he himself was a royal prince of superior status. Laius ordered his charioteer to advance in order to strike Oedipus with his goad. Enraged, Oedipus grabbed the goad, striking and killing Laius plus four of his five retainers, leaving only one to tell the tale.

Upon Laius' death appeared the Sphinx, a lithe monster perched high on the mountain. This creature possessed the body of a dog, the claws of a lion, the tail of a dragon, the wings of a bird and the breasts and head of a woman. The Sphinx began to ravage Thebes, stopping every mountain traveler attempting to enter the city unless they solved her riddle:

"What goes on four feet in the morning, two at midday and three in the evening?"

Whereas the priestess of the Oracle at Delphi revealed a glimpse of the future to her visitors, often concealed in the form of a riddle, the Sphinx, by contrast, killed anyone unable to answer her riddle correctly. The Sphinx either ate or hurled her victims to their death on the rocks below. Until the arrival of Oedipus, the riddle remained unsolved. With no visitors able to enter the city, trade in Thebes had become strangled and the treasury depleted.

Confronted by the Sphinx's riddle, Oedipus responded correctly without hesitation, indicating that it is "mankind" who crawls on four legs in the morning, stands on two in midday and leans on a cane as a third in the twilight of life. Horrified at being outwitted, the Sphinx suffered her own punishment and cast herself to her death on the rocks far below. Consequently Thebes was freed. As reward for saving the city, Oedipus was offered its throne plus the hand of the widow Jocasta. Still unaware of his true origins, Oedipus accepted both honors. He ruled Thebes and married his mother, with whom he multiplied fruitfully. In this manner, Oedipus fulfilled the second part of the oracle.

But the city of Thebes was not finished suffering. Soon it became stricken with a horrible plague and famine that rendered all production barren. Eager to end the affliction, Oedipus once again consulted the oracle. He was told that in order to release Thebes from its current plight, the murderer of Laius must be found. Wanting only what was best for the city, Oedipus relentlessly pursued the quest for the truth. He declared that whenever Laius' murderer was found, the offender would be banished forever from Thebes.

Oedipus called in the blind prophet Tiresias to help, but Tiresias refused to reveal what he knew. Intuiting the truth and dreading the horror of her sins exposed, Jocasta committed suicide by hanging herself. Soon Oedipus discovered that the one he sought was none other than himself. After learning that he had murdered his father and married his mother as predicted, Oedipus was also unable to bear what he saw. He tore a brooch off Jocasta's hanging body and blinded himself. He then faced the consequence that he himself had determined just for Laius' murderer. Subsequently, Oedipus was led into exile by his sister/daughter Antigone.

Here ends the first of Sophocle's tradegies, "King Oedipus." The second and third of this ancient Greek trilogy, "Antigone" and "Oedipus at Colonus," detail Oedipus' and his sister/daughter's extensive wanderings. His tragic insight into unwittingly having committed these crimes of passion brought Oedipus to wisdom. Eventually, he reached a mysterious end in Colonus, near Athens, amidst the utmost respect from his countrymen. Despite his sins, Oedipus' life ended with the blessings of the Gods. To complete one more self-referential loop, his personal insight in-formed the very land itself, as Colonus became an oracular center and source of wisdom for others.

New Twists to an Old Myth

To Freud, the tale of Oedipus was initially conceived in terms of real sexual and aggressive impulses towards real parents, but then his seduction theory was revised and downplayed to the level of fantasy and imaginary impulses. Within Freud's three-part, structural model of the psyche, the Id was the container for unbridled, unconscious, sexual and aggressive impulses; the Super-Ego was a repository for social and societal norms; and the Ego was assigned the difficult task of straddling these two inner, warring factions, while mediating the demands and restrictions of outside reality.

According to Freud, symptoms formed out of the tension between conscious and unconscious factors, including conflicting needs both to repress and express. Among many different kinds of anxiety Freud highlighted, an important symptom was castration anxiety. This was the fear that one's incestuous desire for one's mother would be discovered by the father and punished by him with castration. Both desire for the mother and fear of castration were sources of murderous impulses towards the father. Working through these feelings and symptoms consisted of lifting the repression barrier and thereby gaining insight into the unconscious origins of the conflict.

Note that Freud's developmental model of the psyche was primarily intrapsychic. Because he emphasized the Oedipal complex as a Universal struggle within the internal landscape of all (the adaptation for girls became known as the "Electra" complex, in honor of another famous Greek tragedy), it mattered little how good or bad a child's parenting. Most contemporary psychoanalytic theories, such as object relations (e.g., Klein, 1932), self-psychology (e.g., Kohut, 1971), or intersubjectivity theory (e.g., Stolorow, Brandchaft and Atwood, 1987), have abandoned the importance of the Oedipus myth partly by adopting a more interpersonal focus. Within each of these newer therapies, psychopathology is believed to develop out of real emotional exchanges (or the absence of them) between infants and their caregivers. Symptoms are maintained and altered within the relational context of the analyst/patient dyad.

Prior to these relational theories, near the origins of psychoanalysis, the myth of Oedipus took on an ironic, self-referential twist when it became embodied in real life. Carl Jung, a brilliant follower of Freud, had been earmarked as the "royal son" and "crown prince" slated to inherit Freud's psychoanalytic empire (see Jung, 1961; Kerr, 1995; Monte and Sollod, 2003). The early intimacy and intellectual passion between these two men gave way to great bitterness and struggle surrounding Jung's creative and spiritual ideas. In his autobiography, Jung (1961, p. 150) describes Freud as imploring: "My dear Jung, promise me never to abandon the sexual theory. This is the most essential thing of all. You see, we must make a dogma of it, an unshakable bulwark…against the black tide of mud…of occultism."

For Jung, Freud's topography of the psyche maps only the most superficial level, the "personal unconscious," which contains personal memories and impulses towards specific people. Partly on the basis of a dream, Jung excavated another, even deeper, stratum he called the "collective unconscious." This level had a transpersonal flavor, containing archetypal patterns common in peoples of all cultures and ages.

By acting as if there was room only for what Jung called the "personal unconscious" within the psyche's subterranean zone, Freud appeared compelled to re-enact the Oedipus struggle in real life. He responded to Jung as if to a son attempting to murder his symbolic father. This dynamic was complicated by yet another, even more concrete, level of enactment: both men reputedly competed over the loyalties of the same woman, initially Jung's patient and lover, later Freud's confident, Sabina Speilrein, (see Kerr, 1995).

Freud and Jung acted out the classic Oedipal myth at multiple levels, with Jung displacing Freud both professionally (vanquishing the King) and sexually (stealing the Queen). An explosion ensued when the conflict could no longer be contained or resolved. As a result, the relationship between Freud and Jung was permanently severed. Jung suffered what some believe to be a psychotic break (see Hayman, 1999) and others termed a "creative illness" (see Ellenberger, 1981), from which he recovered to mine the symbolic wealth of his own unconscious.

Jung overcame his symbolic father partly by rejecting the Oedipus myth in favor of Faust's tale. "Jung meant to make a descent into the depths of the soul, there to find the roots of man's being in the symbols of the libido which had been handed down from ancient times, and so to find redemption despite his own genial psychoanalytic pact with the devil" (Kerr, 1995, p. 326). After his break with Freud, Jung self-referentially embodied his own theories about individuation taking the form of the hero's journey. Whereas Jung underscored the sun-hero's motif and role of mythical symbols, mythologist Joseph Campbell (1949/1973) differentiated three phases of the hero's journey: separation (from ordinary consciousness), initiation (into the night journey of the soul) and return (integration back into consciousness and community). This description certainly fits Jung's departure from ordinary sanity, his nightmarish descent into haunting symbols, if not hallucinations, and his professional return to create depth psychology.

Jung and his followers have regarded the Oedipus myth less literally than Freud. In hero mythology, as explicated by one of Jung's most celebrated followers, Eric Neumann (1954/93), to murder the father generally and the King in particular, was seen as symbolic separation from an external source of authority, in order to discover and be initiated into one's own internal source of guidance and wisdom.

Whereas Freud viewed the unconscious primarily in terms of its negative, conflict-ridden potential, Jung recognized the underlying universal and positive potential of the fertile feminine. But in order to uncover this positive side, one first had to differentiate and confront the destructive shadow of the feminine. At the archetypal level, some aspects of the feminine can feel life threatening. To defeat the Sphinx was seen as conquering the Terrible Mother. In her worst incarnation, the Terrible Mother reflected the potential for deprivation and destructive narcissism within the real mother. In some cultures, e.g., the Germanic fairytale of Hansel and Gretel, the Terrible Mother appeared as the Vagina Dentate, or toothed vagina, a cannibalistic allusion not to the Freudian fear of castration by the father, but rather to the Jungian anxiety about emasculation by the mother.

Symbolically, once the dark side of the Terrible Mother was vanquished, her positive potential could be harvested. To have incest and fertilize the mother represented overcoming fear of the feminine, of her dark chaotic womb, in order to tap into riches of the unconscious and bring new life to the psyche. Psychologically we can see how Sphinx and incest fit together for Neumann (1954/93) – the hero killed the Mother's terrible female side so as to liberate her fruitful and bountiful aspect. For Jung, to truly individuate was to rule the kingdom of our own psyche, by overthrowing the father's masculine influence of power, the ultimate authority of consciousness, while fertilizing and pillaging the mother's feminine territory, that of the unconscious. By breaking with Freud and finding his way through his psychosis, Jung killed the King and overcame the Terrible Mother to harvest her symbolism for his own creative development, both in theory and self.

Judging from the drama of real life, both Freud and Jung arrived at their ideas at least partly self-referentially by living them out. Along with affirming Ellenberger's (1981) notion of "creative illness," this coincides with Stolorow's thesis that all significant psychological theory derives from the personal experience and worldview of its originators (Atwood and Stolorow, 1979/1993).

RIDDLE AS PARADOX

As mentioned, in the last several decades, the Freudian interpretation of the Oedipus story largely has been laid aside. With the early advent of feminism, the significance of the tale to a woman's psyche was challenged. With the recognition that sexual abuse was often real and not just fantasy, later feminist thought challenged Freud's early abandonment of his seduction theory. As knowledge about the psychophysiology of the posttraumatic stress condition increased (e.g., Rothschild, 2000), so has clinical interest in "vertical," dissociative splits within the psyche versus the "horizontal" splits maintaining Freud's repression barrier (see Kohut, 1977). Greater relational emphasis among contemporary psychoanalysts shifts interest towards early mother/infant attachment dynamics, as well as here-and-now relations between psychotherapist and patient. Finally, the current climate of multiculturalism disfavors any single theory, especially one universalizing development.

In the spirit of Levi-Strauss, I propose a different way of looking at the Oedipus myth. I aim to harvest meaning primarily by sidestepping narrative content to derive an alternative interpretation both structural and cybernetic in nature. When understood literally, both the "improbable" form the Sphinx embodies plus her impossible-seeming riddle present paradoxes that appear to contradict all known laws of science. Surely no creature on earth can literally walk on four, two and then three limbs during the very same day. With the possible exception of the slime mold, no animal changes its form of locomotion this radically; and not even the slime mold undergoes such complete metamorphosis in the course of a single day.

The Sphinx's riddle presents the type of "ordinary" paradox that science faces all the time. Here paradox is loosely conceptualized as a set of facts that contradicts current scientific theory. Just as Darwin's embodied evolution proceeds in fits and starts (e.g., Gould, 1977), so too does the abstract progression of scientific theory. Kuhn (1962) described the erratic evolution of scientific theory, when resolution of ordinary contradiction leads to abrupt paradigm shifts that offer wider, more inclusive contexts in which to incorporate previously discrepant facts.

Beyond this type of "ordinary" scientific paradox, the Sphinx's riddle was essentially a paradox of self-reference. Its solution – humanity – required deep understanding of the nature of being human, including knowledge of self. In order to know what crawls on four legs in the morning, walks on two in midday and hobbles on three in the evening, Oedipus had to understand the entire human life cycle. He needed to possess intimate familiarity with physical changes in the body, ranging from the dependency of infancy, through the glory of maturity, to the waning powers of old age.

To approach the riddle without self-reference was to look outwards, to use a literal understanding, and to miss a metaphorical interpretation. To approach the riddle with self-reference was to seek knowledge partly by becoming introspective. Oedipus was uniquely positioned to apply the riddle to himself. Almost killed at birth and still physically handicapped, he harbored virtual, vestigial memories of death in life. His limp and cane were whispers of a helpless past and harbingers of a shattered future.

Self-referentially Oedipus' own life trajectory showed the same three parts as the Sphinx's riddle. Through the kindness of others Oedipus survived the traumatized helplessness of infancy. In his prime, he proved more than able to stand on his own two feet – strong enough to kill a king, clever enough to slay the proverbial monster, and potent enough marry a queen and spawn a covey of offspring. Ironically in the case of our tragic hero, it was Oedipus' very in-sight into his own origins that led to the loss of his kingdom and wife/mother, leaving him to hobble around blindly in old age, leaning on his cane, and dependent upon the goodness of others, primarily his daughter/sister, Antigone.

The namesake and body memories of Oedipus connected him with chance and destiny, past and future, infancy and old age. Recall that the name Oedipus means both "swollen foot" and "know-where." Feder (1974/1988) analyzed the Oedipus myth in terms of the clinical reality of adoption trauma. Like many adopted children, Oedipus was relentlessly driven to seek his own origins in order to "know where" he came from both genetically and socially.

Taking this approach a step further, we can see the impact of early physical abuse – attempted infanticide – on the neurobiology of different memory systems. Oedipus "knows where" he came from implicitly in his body due to his "swollen foot," even while ignorant of traumatic origins explicitly in his mind. This kind of *implicit memory* has gained much attention in recent clinical lore (e.g., Rothschild, 2000; Siegel, 2001). In early infant development, *implicit memory* is the first kind to develop. It is believed to help tune ongoing perception and emotional self-regulation in the nonverbal context of relationships with others. In this way contingent versus non-contingent responses of caretakers become hardwired into the brain and body via particular neural pathways. While alluded to by others, e.g., Ornstein (1973), Allan Schore (2001) specifically proposes that implicit memory exists within the right, nonverbal, hemisphere of the human cerebral cortex to constitute the biological substrate for Freud's unconscious instincts and memories. Although hotly contested, neurobiological evidence mounts for Freud's repression barrier as hardwired into the brain (e.g., Solms, 2004).

Schore proposed a vertical model of the psyche, where the conscious, verbal, mind is localized in the left hemisphere of the brain, while the unconscious and body memory is mediated by the nonverbal right hemisphere (for most right handed people). The hemispheres of the brain and these different modes of processing are conjoined as well as separated by the corpus callosum. Early trauma plus his secret origins caused a haunting and widening of the gap between what Oedipus' body versus his mind knew. Oedipus' implicit memory of his early abandonment and abuse became the invisible thread that provided deep continuity despite abrupt life changes. His implicit memory offered a clue to the commonality beneath the apparent disparity in the Sphinx's three-part riddle.

Structurally, to solve the riddle became equivalent to Oedipus' self-referential quest for explicit memory of his own origins. This interpretation meshes with anthropologist Lévi-Strauss' (1977) emphasis on structural similarities within and between myths, plus the near universal concern with human origins. It also dovetails with Bion's (1983, p. 46) self-referential understanding of the Sphinx's riddle as "man's curiosity turned upon himself." In the form of self-conscious examination of the personality by the personality, Bion uses the Oedipus myth to illuminate ancient origins of psychoanalytic investigation.

Metaphorical Thinking and Cognitive Development

In order to solve both the riddle of the Sphinx as well as that of his own origins, Oedipus had to delve beneath the concrete level of surface appearances. Here he'd lived happily, but in ignorance, as children and innocents are reputed to do. Ignorance might have been bliss, but it didn't necessarily lead to maturity. Prior to Oedipus solving the riddle, humankind lived in an immature state, an idea supported by the work of Julian Jaynes (1976). Writing about the "bicameral mind," Jaynes speculated that ancient humanity hallucinated gods as living in their midst. Here myths were concretely embodied, serving as externals sources of authority before such executive functions became internalized within the cerebral cortex of the modern psyche, including our increased capacities for self-reflection, inner guidance and self-control.

The Sphinx's riddle was a self-referential mirror reflecting and later enabling explicit memory and knowledge of Oedipus' traumatic origins. Upon successfully answering the riddle, Oedipus bridged the earlier, developmental territory of the "right mind" with the evolutionarily and developmentally later left-brain (see Schore, 2001). In the process, Oedipus healed and matured on many levels. Not only did he address his castration fears by conquering the Terrible Mother in the form of the Sphinx after killing the Terrible Father, but also and perhaps more significantly, Oedipus made the leap from concrete to metaphorical thinking. By understanding "morning," "midday" and "evening" as stages of life, he demonstrated creativity and mental flexibility characteristic of internal complexity. Cognitive psychologists Lakoff and Johnson (1980) have suggested that metaphor serves as the basis for all abstract thinking. More recently, Lakoff and Johnson (1999) argued that metaphor forms part of the implicit memory of the cognitive unconscious, where its immediate conceptual mapping is hard-wired into the brain.

The leap from concrete to metaphorical thinking not only was an important developmental step in the history of consciousness, but it also can be understood within the historical trajectory of the individual. Here Piaget's developmental epistemology (e.g., Flavell, 1963) becomes relevant. Though details are still disputed, overall Piaget's theory has remained one of the most important and universal accounts of intellectual development to date (see Sternberg, 1990). Using careful observation and empirical studies, Piaget mapped the shift from a sensorimotor period of infancy, through the pre- and concrete operations of early childhood, into a formal operations stage of later childhood characterizing the adult, "mature" mind. Piaget's hallmark of maturity involved freedom from the particulars of concrete situations, granting cognitive flexibility necessary both for abstract and metaphorical thinking.

Self-Reference and Universal Turing Machines

So far I have suggested that self-reference is central to a metaphorical solution of the Sphinx's riddle. But self-reference also proves to be an essential part of cybernetics, the sciences of information. A computational model views the human psyche as a recursive system, where present behavior depends upon how it has processed its past behavior. Within abstract machines, different computational powers depend deterministically upon a system's retrospective access to memory.

In computational science, power is ranked according to "Chomsky's hierarchy." At the bottom of the hierarchy lies the finite state automaton. This machine possesses only *implicit* memory for its current state. In the middle lies the push-down automaton. This machine possesses *explicit* memory, but with only temporary access to the past. At the top of Chomsky's hierarchy lies the Universal Turing Machine. This abstract machine possesses unrestricted, permanent and explicit memory for all past states.

Cyberneticist Ron Eglash (1999) provides a text analogy to contrast these differences: The least powerful machine is like a person who accomplishes all tasks instinctively, without the use of any books; in the middle is a person limited by books removed once they've been read; at the top is a person who collects and recollects all books read, in any order. The power of the Universal Turing Machine at the top is its capacity to recognize all computable functions.

The point at which complete memory of past actions is achieved marks a critical shift in computational power. It is the point when full self-reference is achieved, which brings about the second-order, cybernetic capacity of a system to analyze its own programs. My reading of the Oedipus myth illustrates this very same point – that powerful instant when full access to memory dovetailed with self-reference to signal another step in the "complexification" of human consciousness.

The Riddle as Mirror

Just as the Sphinx presented a paradigm of self-reference to hold a mirror up to Oedipus, the myth of Oedipus also holds a mirror up to us as witnesses. The story of Oedipus reflects our own stories in yet another self-referential loop. Like Oedipus, each one of us is a riddle to him or herself. The story rocks generation after generation so powerfully partly because of this self-referential quality, which forces each one of us to reflect upon our own lives mythically.

Throughout the tale, there is dynamic tension between knowing and not-knowing – in Oedipus and in us. Oedipus starts out naïvely not-knowing who he is or where he came from. We start out knowing who Oedipus really is, but blissfully unaware of the truth in ourselves. By the end of the tale, the situation reverses: Oedipus solves all three riddles, that of the Oracle of Delphi, that of the Sphinx and that of his origins, while ironically, we participant/observers are left not-knowing. We harbor a gnawing feeling of uncertainty – almost as if another riddle has invisibly materialized, as if we face the very Sphinx herself, whose enigma must be answered upon threat of our own death.

Eglash (1999) notes that the power of the Universal Turing Machine lies in its ability *not to know* how many transformations, or applications of an algorithm a system would need ahead of time, before the program could be terminated. *Paradoxically, to achieve full uncertainty about the future and its relationship to the past is symptomatic of increasing computational power.* This kind of fundamental uncertainty is evident collectively within the modern sciences and mathematics of chaos theory, stochastic analyses, and various forms of indeterminacy. For example, Heisenberg's uncertainty principle states the impossibility of precisely determining both a quantum particle's speed as well as its location at the same time. Meanwhile, chaos theory warns of the impossibility of precisely predicting the long-term future of highly complex systems, no matter how precise our formulas or capacity to model their past behavior.

Experientially, we must deal with fundamental uncertainty with respect to the riddle of our own lives, leaving us ultimately responsible to glean meaning from this self-reflective search. The Oedipus myth presents a self-referential mirror through which each one of us individually enters the modern stage of self-reflective consciousness. Capabilities for full memory, to consider the past and future, to contemplate death, to confront paradox, to self-reflect and to consider self-reference all represent critical levels of inner complexity that separate human from animal intelligence, the infant from the mature individual, plus the weakest from the most powerful computing machines.

I end this paper by speculating how this complex state of full self-reference serves as a prerequisite to a fully self-actualized human being. To have thorough access to memory for the past plus the cognitive flexibility not to have to know the future represents a state of high integration between left and right brain hemispheres, between body and mind, and between implicit, procedural memory versus explicit memory for events and facts. Such integration maximizes our potential for spontaneous response and creative self-expression that is the hallmark of successful individuation.

Furthermore, I argue that this complex state of "good-enough" self-reflective awareness is necessary to break the tragic intergenerational chain of fate and trauma symbolized by Greek tragedy in general and the Oedipus myth in particular. At the heart of the Oedipus myth lies the observation, echoed by a Greek chorus, that those born into abuse unwittingly grow up to become abusers. Laius' unsuccessful attempt to kill his son all but sealed Oedipus' fate to escalate this loop of violence by successfully killing his father.

The only way out of the fatalistic tragedy of abusers begetting abusers is to possess enough insight to unearth violent instincts before the deed is done, to exert sufficient self-control to resist and transcend such instincts, plus to tell a cohesive, self-reflective narrative. Multigenerational, prospective research within the field of attachment (e.g., Siegel, 1999) suggests that the best predictor of healthy, secure attachment in children remains the capacity for their parents to tell a cohesive narrative about their early childhood. It matters little whether the quality of this narrative is idyllic or horrific. What counts instead is whether parents possess the self-reflective insight to hold onto specific memories concerning their origins, which can be cohesively woven into the fabric of current life without disruption. This kind of self-referential reflection carries the full computational power of Universal Turing Machine. This provides the necessary mental faculties to break intergenerational chains of emotional and physical abuse. It also allows for creative self-actualization, without a pre-determined script, set upon the stage of an open future.

REFERENCES

Aristotle (1982, trans. by W. Fyfe). *Poetics*, vol. 23, Loeb Classical Library. Cambridge, MA: Harvard University Press.

Atwood, G. and Stolorow, R., (1979/1993). *Faces in a Cloud: Intersubjectivity in Personality Theory*. Northvale, New Jersey: Jason Aronson.

Bion, W. (1983). *Elements of Psycho-Analysis*. Northvale, NJ: Jason Aronson.

Campbell, J. (1949/1973). *The Hero With a Thousand Faces*. Princeton, NJ: Bollingen Series, Princeton University.

Eglash, R. (1999). *African Fractals: Modern Computing and Indigenous Design*. New Jersey: Rutgers University Press.

Ellenberger, H. (1981). *The Discovery of the Unconscious*. New York: Basic Books.

Feder, L. (1974). Adoption Trauma: Oedipus Myth/Clinical Reality. In *The Oedipus Papers*, Pollock, G. and Ross, J., eds. (1988). Madison, CT: International Universities Press.

Flavell, J. H. (1963). *The Developmental Psychology of Jean Piaget*. New York: Van Nostrand.

Freud, S. (1900). *The Interpretation of Dreams* (Trans. J. Strachey). New York: Basic Books.

Gould, S. J., and Eldredge, N. 1977. Punctuated equilibria: the tempo and mode of evolution reconsidered. *Paleobiology*, 3, 115-151.

Hayman, D. (1999). *The Life of Jung*. New York: W.W. Norton.

Heims, S. (1991). *The Cybernetics Group*. Cambridge, MA: The MIT Press.

Jaynes, J. (1976). *The Origin of Consciousness in the Breakdown of the Bicameral Mind*. Boston: Houghton Mifflin.

Jung, C. (1956). Symbols of Transformation. In *Collected Works*. London: Routledge and Kegan Paul.

Jung, C. (1961). *Memories, Dreams , Reflections*. New York: Random House.

Kerr, J. (1995). *A Most Dangerous Method*. New York: Vintage Books/Random House.

Klein, M. (1932). *The Psycho-Analysis of Children*. London: Hogarth.

Kohut, H. (1971). *The Analysis of the Self*. New York: International Universities Press.

Kohut, H. (1977).*The Restoration of the Self*. New York: International Universities Press.

Kuhn, T. (1962). *The Structure of Scientific Revolutions*. Chicago, Ill: University of Chicago Press.

Lakoff, G. and Johnson, M. (1980). *Metaphors We Live By*. Chicago: University of Chicago Press.

Lakoff, G. and Johnson, M. (1999). *Philosophy in the Flesh: The Embodied Mind and Its Challenge to Western Thought*. New York: Basic Books.

Lévi-Strauss, C. (1977). *Structural Anthropology 1* (trans. Jacobson, C. and Schoepf, B. G.). Harmondsworth: Penguin.

Masson, J. (1984). *The Assault on Truth: Freud's Suppression of the Seduction Theory*. Horizon Book Promotions

Monte, C. and Sollod, R. (2003). *Beneath the Mask: An Introduction to Theories of Personality*. New York: John Wiley and Sons.

Nietzsche, F. (1871/1999) *The Birth of Tragedy and other Writings (Cambridge Texts in the History of Philosophy).* Cambridge, UK: Cambridge University Press.

Neumann, E. (1954/1993). *The Origins and History of Consciousness*. Princeton, NJ: Princeton.

Ornstein, R. (Ed., 1973). *The Nature of Human Consciousness*. San Francisco: W.H. Freeman.

Pollock, G. and Ross, J. (1988). *The Oedipus Papers*. Madison, CT: International Universities Press.

Ricoeur, P. (1970). *Freud and Philosophy*. Cambridge, MA: Yale University Press.

Rothschild, B. (2000). *The Body Remembers: The Psychophysiology of Trauma and Trauma Treatment*. New York: W. W. Norton.

Schore, A. (2001). Minds in the making: Attachment, the self-organizing brain, and developmentally-oriented psychoanalytic psychotherapy. *British Journal of Psychotherapy*, 17, 3, pp. 299-328,

Siegel, D. (2001). Memory: An overview, with emphasis on developmental, interpersonal, and neurobiological aspects. *Journal of the Academy of Child and Adolescent Psychiatry*, 40, 9, pp. 997-1011.

Solms, M (2004). Freud returns. *Scientific American*, May, pp. 83-89.

Sternberg, R. (1990). *Metaphors of Mind: Conceptions of the Nature of Intelligence*. New York: Cambridge University Press.

Stolorow, R., Brandchaft, B., and Atwood, G. (1987). *Psychoanalytic Treatment: An Intersubjective Approach.* Hillsdale, NJ: The Analytic Press.

In: Progress in Chaos Complexity Research
Editors: Franco F. Orsucci and Nicoletta Sala
ISBN: 978-1-60456-375-7

Chapter 7

COMPARISON OF EMPIRICAL MODE DECOMPOSITION AND WAVELET APPROACH FOR THE ANALYSIS OF TIME SCALE SYNCHRONIZATION

Dibakar Ghosh[a,b,*] ***and A. Roy Chowdhury***[b,2]
[a]Department of Mathematics, Dinabandhu Andrews College,
Garia, Calcutta-700 084, India
[b]High Energy Physics Division, Department of Physics,
Jadavpur University, Calcutta- 700 032, India

ABSTRACT

In this letter, we address the time scale synchronization between two different chaotic systems from the view point of empirical mode decomposition and the results are compared with those obtained using wavelet theory. In empirical mode decomposition method, we decompose a time series in distinct oscillation modes which may display a time varying spectrum. In this process it was observed that the transition of non synchronized, phase synchronization, lag synchronization and complete synchronization occurs for different coupling parameter values. The quantitative measure of synchronization for empirical mode decomposition and wavelet approach has been proposed. It has been observed that due to the presence of a scaling factor in the wavelet approach it has more flexibility for application.

1. INTRODUCTION

Synchronization phenomenon was discovered by Huygens in the year 1665, when he found that two very weakly coupled pendulum clocks, hanging from the same beam, become phase synchronization. Synchronization in nonlinear dynamical systems is one of the most important aspect which is now studied both for its importance in theory and experiment.

Experimental implications of synchronization actually crosses boundaries of different domains of science and technology. From the subject of communication system to that of epilepsy in brain or cardiac properties of heart all can be studied with the help of chaotic synchronization [1]. At the present moment there exists four types of synchronization, namely -- complete, lag, generalized and phase synchronization [1, 2].

Complete synchronization[CS] is characterized by the convergence of two chaotic trajectories i.e. y(t)=x(t). Generalized synchronization is defined as the presence of some functional relation between the response state x(t) and drive state y(t). Phase synchronization[PS] means entrainment of phases of chaotic oscillators, $n\Phi_x - m\Phi_y$ =constant where n and m are integers, while their amplitudes remain chaotic and uncorrelated. Lag synchronization[LS] for the first time was introduced by Rosenblum et al.[2] under certain approximations in studying synchronization between bidirectional coupled systems described by the ordinary differential equation with parameter mismatches i.e. $y(t)=x(t-\tau)$ with positive τ.

Several group of researchers have advocated various effective methods for all of the above processes. But the case of phase synchronization is always a bit different, essentially due to inherent difficulty of definition of phase in a chaotic system. Various authors [1, 3] have proposed completely different approaches for the analysis of phase of nonlinear chaotic systems. If the signal possesses a multicomponent spectrum i.e. it does not display a single oscillation at a unique frequency, the usual methods for detection of phase are failed. Thus there is no unique method to determine the phase in complex chaotic oscillators, and different definition of the phase can be found [1]. Moreover, coupling between different chaotic systems can cause significant deformation of their attractors even in the case of simple chaotic oscillators, which also makes impossible defining the phase and studying PS on the basis of simple physical intuitions. Here, we mention a very general approach suggested by Huang [4] which goes by the name of Empirical mode decomposition [EMD]. In this approach the basic idea is to separate the various frequencies in different intrinsic mode functions [IMF]. Each of these having a well defined frequency. And for the study of synchronization, it is easy to compare the various IMF's, for the two signals under consideration.

In this communication we study the time scale synchronization between two different chaotic systems with several spectral components. We analyze different route to synchronization through EMD and wavelet approach and compare their results. The quantitative measure of synchronization for empirical mode decomposition and wavelet approach has been proposed.

2. Models

The systems under consideration is that of two model from ecological systems. They are quite similar and can be expressed as –Model-I Model-II

* E-mail: drghosh_chaos@yahoo.com

Model 1

$$\frac{dx}{dt} = ax\left(1 - \frac{x}{k}\right) - \frac{bxy}{d_1 + x}$$

$$\frac{dy}{dt} = -cy + \delta xy - \frac{\eta yz}{d_z + y}$$

$$\frac{dz}{dt} = -\varphi z + \gamma yz$$

Model 2

$$\frac{du}{dt} = au\left(1 - \frac{x}{k_1}\right) - \frac{a_2 uv}{b_2 + u}$$

$$\frac{du}{dt} = -c_2 \frac{a_2 uv}{b_2 + u} - \delta_2 v - \frac{a_3 vw}{b_3 + v}$$

$$\frac{dw}{dt} = -\delta_3 w + c_3 \frac{a_3 vw}{b_3 + v}$$

where systems parameters are all positive.

3. Empirical Mode Decomposition

Phase synchronization of chaotic systems is the appearance of a certain relation between the phases of the interacting systems while the amplitudes remain uncorrelated. There is no unambiguous and general definition of phase for any chaotic system. Three well known approaches to phase determination are: (i) Sometimes the projection of the attractor on some planes looks similar to a smeared limit cycle. The trajectory thus rotates about a point and the Poincare section can be chosen in a proper way. With the help of the Poincare map, a phase can be defined, attributing to each rotation the 2π increase [1]: $\varphi_M = 2\pi \frac{t - t_n}{t_{n+1} - t_n} + 2\pi n$, $t_n \le t < t_n$ where t_n is the time of the n-th crossing of the secant surface. (ii) If the above mentioned projection is found, the phase can be defined as the angle between the projection of the phase point on the plane and a given direction on the plane [2, 3] $\varphi_M = \tan^{-1}\left(\frac{y}{x}\right)$ (iii) A different way to define phase is based on Hilbert transform and was originally introduced by Gabor [5] which unambiguously gives the instantaneous phase and amplitude for an arbitrary signal s(t). The analytic signal ζ (t) is a complex function of time [1, 3] defined as $\zeta(t) = s(t) + i\,\tilde{s}(t) = A\, e^{i\varphi_H(t)}$ where the function $\tilde{s}$ (t) is a Hilbert transform of s(t). The instantaneous amplitude A(t) and $\varphi_H(t)$ the instantaneous phase of the signal s(t) are thus uniquely calculated.

The phase variable $\varphi(t)$ can be easily estimated from a scalar time series x(t). But the problem arises when the signal possesses a multicomponent or a time varying spectrum. In that case, its trajectory on the complex plane may show multiple centers of rotation and an instantaneous phase cannot be defined easily [6, 7]. The power spectrum of the signal indicates multiple Fourier modes, which creates a difficulty in the estimation of phase. To overcome this difficulty we use the algorithm of EMD. Explicit procedure of EMD for a given signal x(t) can be summarized [8] as follows.

(i) All extrema of x(t) should be identified.
(ii) Interpolation should be done between the minima (respectively maxima) ending up with some envelope e_{min} (t) (respectively e_{max} (t)).
(iii) The mean m(t)=(e_{min} (t) + e_{max} (t))/2 should be computed.
(iv) The detail d(t)=x(t)-m(t) should be extracted.
(v) The residual m(t) should be iterated.

In practice, a *sifting process* refines the above procedure which amounts to first iterating steps (i) to (iv) upon the detail signal d(t), until the latter can be considered as zero-mean according to some stopping criterion.. Once this is achieved, The detail can be referred to as an *Intrinsic Mode Function* (IMF). The corresponding residual is computed and step (v) applies. The number of extrema decreases while going from one residual to the next and the whole decomposition is completed within a finite number of modes. EMD ensures that the complex plane of $C_j(t)$ rotate around a unique rotation center and once this process is achieved, the resulting signal can be considered as a proper rotation mode and the phase can be defined.

4. Time Scale Synchronization between Two Different Chaotic Systems via EMD Method: From Phase to Lag Synchronization

Model I is unidirectionally coupled with Model II and the third equation of the first model gets modified as:-

$$\dot{z} = -\varphi z + \gamma y z + s_1 (w-z)$$

where s_1 is the coupling coefficient. The two systems are coupled when they are in a chaotic state. The Model I is in a chaotic state for a=3.2, while the others parameters are k=50, b=1, d_1=10, c=1, δ =0.05, η =1, d_2 =10.5, φ=1, γ =0.05. and the Model II is in a chaotic state for α =3.1, k=1, a_2=1.66, d_2=0.33, c_2=1, δ_2=0.27, a_3=0.05, b_3=0.5, c_3=1, δ_3=0.01. When the coupling parameter s_1 is equal to zero, the chaotic attractors and Fourier spectrum of the coupled system are shown in figure(1a, 1b) and figure(1c, 1d) respectively. These Fourier spectra shows multiple Fourier modes i.e. the chaotic signal contains more than one main frequency. The attractors in figure (1a) and (1c) does not show any unique center of rotation. Thus the procedure to calculate phase by method (i) and method (ii) fails. The trajectory of their analytic signal in the complex plane also does not show any unique centre of rotation which is depicted in Figure (1e) and (1f) respectively and thus the third method (i.e. Hilbert transform) is also ruled out.

We have decomposed the original chaotic signals z(t) and w(t) as

$$z(t)=\Sigma_{j=1}^{M} C_j (t) + R(t) \quad \text{and} \quad w(t)=\Sigma_{j=1}^{N} C'_j (t) + R'(t)$$

where R(t) and R'(t) are residuals of the signals z(t) and w(t) respectively. The functions $C_j(t)$ (and $C_j'(t)$ are nearly orthogonal to each other. Thus each mode generates a proper rotation on the complex plane with the analytic signal

$$C_j(t) = A_j(t)e^{i\varphi j(t)} \text{ and } C_j'(t) = A'_j(t)e^{i\varphi' j(t)}$$

and the two phases φ_j and φ'_j of the signals $C_j(t)$ and $C_j'(t)$ respectively is obtained.

Thus the phases φ_j and φ'_j are obtained by using Hilbert transform of each $C_j(t)$ and $C_j'(t)$ respectively and the frequencies ω_j and ω'_j are obtained by averaging the instantaneous frequencies $\frac{d\varphi_j}{dt}$ and $\frac{d\varphi'_j}{dt}$ respectively, separately for each mode.

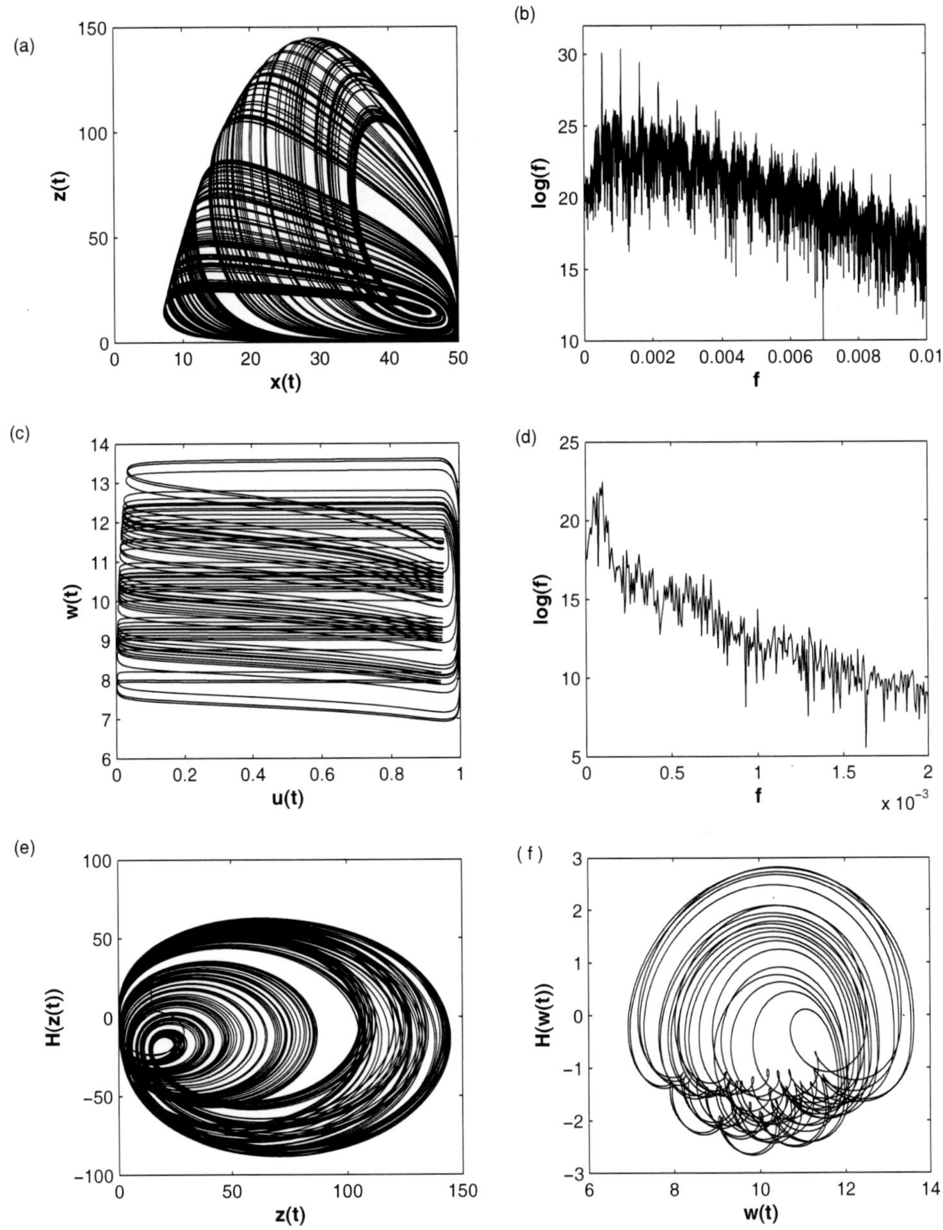

Figure 1. (a) Phase plot of system I, (b) Fourier spectrum of system I, (c) Phase plot of system II, (d) Fourier spectrum of system II, (e) Trajectory of the analytic signal z(t) on the complex plane, (f) Trajectory of the analytic signal w(t) on the complex plane (see the multiple centers of rotation).

The instantaneous frequencies of each mode can vary with time and the fast oscillations present in the signal are in general extracted into the lower and the slow oscillations into the higher modes so that $\omega_0 > \omega_1 > \omega_2 > \ldots > \omega_M$ and $\omega'_0 > \omega'_1 > \omega'_2 > \ldots > \omega'_N$. Moreover, the mode amplitudes usually decay fast with j so that the signal can be decomposed into a small number of empirical modes. In this context one should note that the phase synchronization between the drive and response systems can always be characterized either as a phase locking or a weaker condition of frequency locking. One should note that strictly speaking phase locking and mean frequency locking are two independent condition to characterized phase synchronization [1, 3]. In the synchronized state the phase difference between the oscillators is bounded and the frequency difference is zero or at least close to zero. By using the EMD method, the transition to phase synchronization is basically analyzed as a process of merging of the frequencies of the corresponding modes and the phase difference is bounded as the coupling strength is increased. Further, when synchronization is set, different types of phase interactions may simultaneously arise at specific time scales.

For different values of the coupling parameter $s_1$the IMF are obtained for the signal z(t), viz., 7 IMF's and a residue for $s_1=1.5$, 5 IMF's and a residue for $s_1=3$, with further increase of coupling parameter we get 5 IMF's and a residue for $s_1=50$. From this we concluded that the number of IMF's decreases with the increase of coupling parameter s_1 which indicates that the system has entered into a ordered state from a chaotic one. To obtain the phase of the system it is necessary to observe whether the phase plot shows an unique center of rotation or not. For this purpose, the trajectory of the intrinsic mode functions $C_j(t)$ and $C_j'(t)$ in the complex plane are plotted distinctly and shows the center of rotation as per our requirement.

The behavior of phase difference $[\Delta(\varphi)]$ at different intrinsic time scales for different coupling strength $s_1=1.5$ and $s_1=50$ are shown in Figure (2a) and Figure (2b) respectively. We observed that in figure (2a) for low coupling strength $s_1=1.5$ the phase difference between different intrinsic mode are not bounded. The phase difference φ_j - φ'_j are not bounded for almost all intrinsic time scales. For further increases of parameter (i.e. $s_1=50$) the intrinsic time scales of the first chaotic oscillator becomes correlated with the other intrinsic modes of the second oscillator and the phase synchronization[PS] occurs [in figure (2b)]. With further increase of coupling (i.e. $s_1=100$) the transition of PS to lag synchronization[LS] occurs. The LS between oscillators means that all intrinsic modes are correlated. From the condition of LS we have $z(t-\tau) \cong w(t)$ and therefore $\varphi_j(t-\tau) \cong \varphi'_j(t)$ where τ is the time lag. For further increase of coupling parameter ($s_1=350$) leads to the decreasing of the time lag and transition of LS to complete synchronization occurs.

An very interesting phenomena observed when the above two models I and II are coupled bidirectionally, the resulting coupling system read:

$$\frac{dx}{dt} = ax\left(1-\frac{x}{k}\right) - \frac{bxy}{d_1+x} \qquad \frac{du}{dt} = au\left(1-\frac{x}{k_1}\right) - \frac{a_2uv}{b_2+u}$$

$$\frac{dy}{dt} = -cy + \delta xy - \frac{\eta yz}{d_2+y} \qquad \frac{du}{dt} = -c_2\frac{a_2uv}{b_2+u} - \delta 2\,v - \frac{a_3vw}{b_3+v}$$

$$\frac{dz}{dt} = -\varphi z + \gamma yz + s'_1(w-z) \qquad \frac{dw}{dt} = -\delta 3\,w + c_3\frac{a_3vw}{b_3+v} + s'_1(z-w)$$

where s'_1 is coupling strength. For small coupling parameter value the instantaneous phase difference $\Delta(\varphi)$ diffuse at all intrinsic time scales i.e. the time scales are unsynchronized state.

For increase of coupling strength $s'_1 = 6.5$ the phase difference of all intrinsic time scales between two oscillators are bounded [Figure 3]. With small increase of coupling strength $s'_1=6.6$ the LS occurs and at this position the phase difference between modes are not equal to zero but very close to zero. The time lag decreases with increase of coupling strength and complete synchronization[CS] occurs at $s'_1=20$.

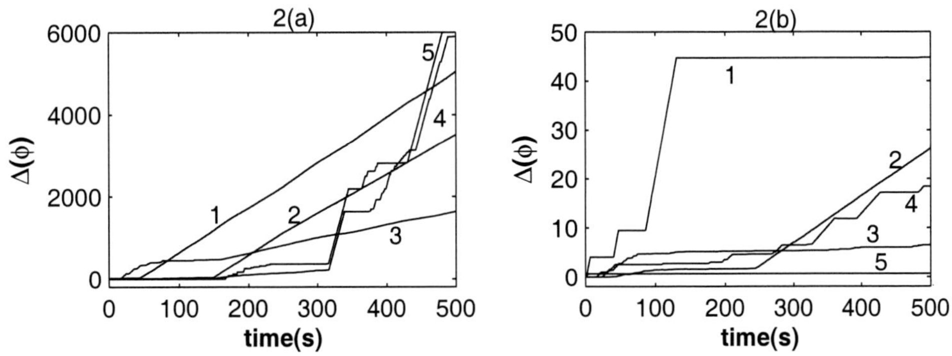

Figure 2. Representation of the phase difference of the system I and II when unidirectionally coupled together at coupling parameter (a) $s_1=1.5$, (b) $s_1=50$, (1 represent the phase difference of the corresponding C_1 and C'_1 IMF's, 2 represent the phase difference of the corresponding C_2 and C'_2 IMF's and so on).

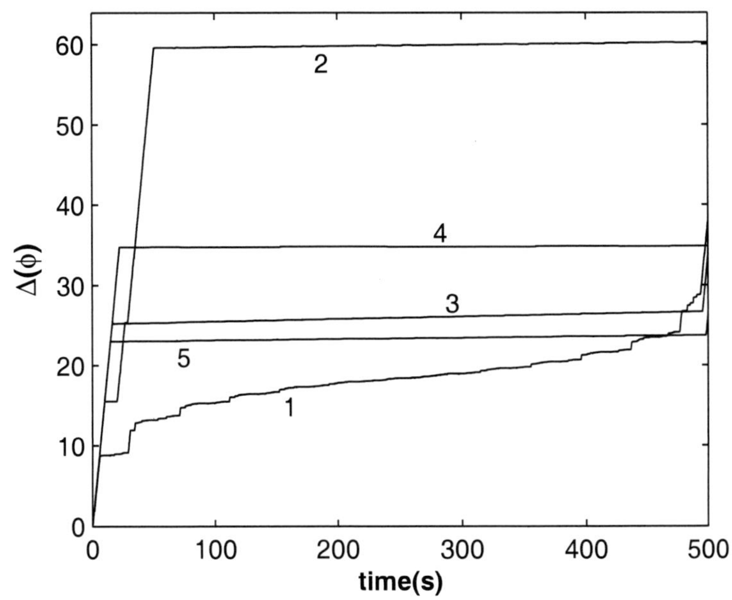

Figure 3. Representation of the phase difference of the system I and II when bidirectionally coupled together at coupling parameter $s_1=6.5$, (1 represent the phase difference of the corresponding C_1 and C'_1 IMF's, 2 represent the phase difference of the corresponding C_2 and C'_3IMF's, 3 represent the phase difference of the corresponding C_5 and C'_3 IMF's , 4 represent the phase difference of the corresponding C_7 and C'_5 IMF's, 5 represent the phase difference of the corresponding C_3 and C'_4 IMF's).

At this position the phase difference between two intrinsic time scales zero i.e. $\varphi_j(t) \cong \varphi'_j(t)$. It is observed from above that the transition of nonsynchronized, PS, LS, CS are occurs depending upon the strength of coupling.

5. MEASURE OF SYNCHRONIZATION FOR EMD

In the previous section we have discussed the transition of synchronization (non-synchronization, PS, LS, CS) occurs depending upon the coupling strength. Therefore, the measure of synchronization be introduced. This measure σ can be defined as

$$\sigma = \frac{\langle[\varphi_j(t) - \varphi'_j(t)]^2\rangle}{\|\varphi_j(t)\|^2 \|\varphi'_j(t)\|^2}$$

where <> represents averages over the phase difference $\varphi_j(t) - \varphi'_j(t)$ and $\|\varphi(t)\|$ are norm of $\varphi(t)$. This measure σ is near to 1 for non-synchronized oscillations and 0 for complete synchronized regime. For the LS this measure near to 0. If the phase synchronization regime is observed, it takes value between 0 and 1 depending upon synchronized time scales. So the synchronization measure σ does not allow to distinguished the non-synchronized and synchronized oscillation but characterize quantitatively the degree of intrinsic time scales synchronization. The variation of σ with changes of bidirectional coupling strength s'_1 is shown in Figure 4. From this figure it is observed that for coupling s'_1=0.3 the system are in unsynchronized state and for s'_1=6.6 there is a lag synchronization state. For s'_1=20 the complete synchronization occurs and at this point σ.

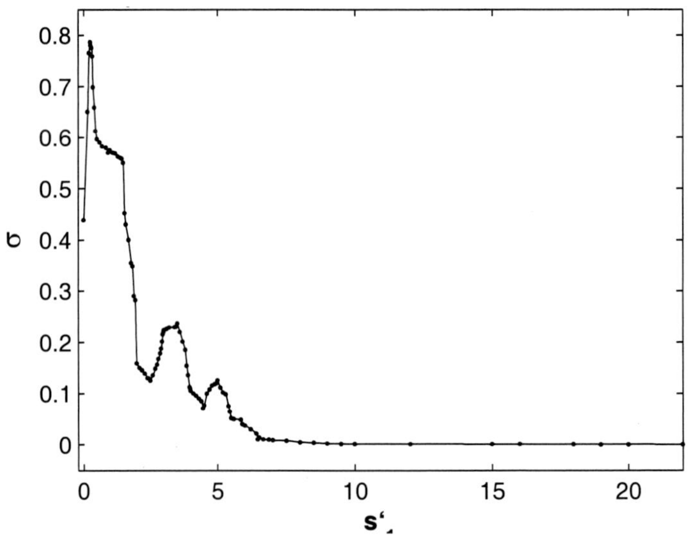

Figure 4. The dependence of the synchronization measure σ for different coupling strength s'_1.

6. WAVELET APPROACH (WA)

The alternative approach for the analysis of phase of a complicated time series is the wavelet transform. Several people have already worked on this approach [9, 10]. To elaborate the basis of wavelet technique, consider a time series x(t). The behavior of such systems can be characterized by a continuous phase set define wavelet transform of the chaotic signal x(t);

$$W(s_0, t_0) = \int_{-\infty}^{\infty} x(t)\psi s_0, t_0 dt$$

where the asterisk means complex conjugation and$_{s\ 0}$

$$\psi s_0, t_0(t) = \frac{1}{\sqrt{s_0}}\psi_0\left(\frac{t - t_0}{s_0}\right)$$

is the wavelet function obtained from the mother wavelet $\varphi s_0,\ t_0\ ^{(t)}$. The time scale s_0 determines the width of $\psi_0(t)$, where t_0 stands for time shift of the wavelet function. Here it should be noted that the time scale s_0 is replaced by the frequency of Fourier transform and can be considered as the quantity inverse to it. Usually their are some very standard and basic wavelet also used in which we use the Morlet wavelet also used in references [10]. It is given as;

$$\psi_0(\eta) = \frac{1}{\sqrt[4]{\pi}} \exp(i\omega_0\eta) \exp\left(-\frac{\eta^2}{2}\right)$$

The wavelet parameter $\omega_0 = \dot{2}$ ensures the relation $s = \frac{1}{f}$ between the time scale s and frequency f of Fourier transform. One then considers

$$W(s_0, t_0) = |W(s_0, t_0)|\, e^{i\varphi s0(t0)}$$

which determines the wavelet surface characterizing the behavior of the system for every time scale s_0 at any time t_0. The magnitude of $W(s_0, t_0)$ represents the relative presence and magnitude of the corresponding time scale s_0 at t_0. Usually it is very standard to consider the integral energy distribution over all time as

$$\langle E(s_0)\rangle = \int |W(s_0, t)|^2\, dt_0$$

The phase $\varphi t_0\ (t) = \arg W(s_0, t_0)$ also proves to be naturally defined for time scale s_0. So that the behaviour of each time scale s_0. So that the behaviour of each time scale s_0 can be identified using the phase $\varphi s_0(t)$. We now apply this idea to the two time series obtained from the two chaotic systems. It is observed that the time scales accounting for the greatest fraction of the wavelet spectrum energy $\langle E(s_0)\rangle$ are obviously synchronized first. For other time scales there is no synchronization. Actually we should have $|\varphi s_1^{(t)} - \varphi s_2^{(t)}| <$ constant, for some s, leading to phase locking in the situation of phase synchronization. In figure (5a) we show the variation of <E(s_0)> against s_0, for both the system. It is interesting to observed that the two almost totally overlap having a peak at the scale s_0 =14. In the adjoining figure, we display for the various s_0 and their corresponding phase differences are shown in figure (5b). It is clear, that the phase difference $\varphi s_1(t)^{-}\ \varphi s_2(t)$ is not bounded for all time scale. From this figure its observe that at s_0 =14 the difference remains zero for all time. For other time scales the phases are not locked.

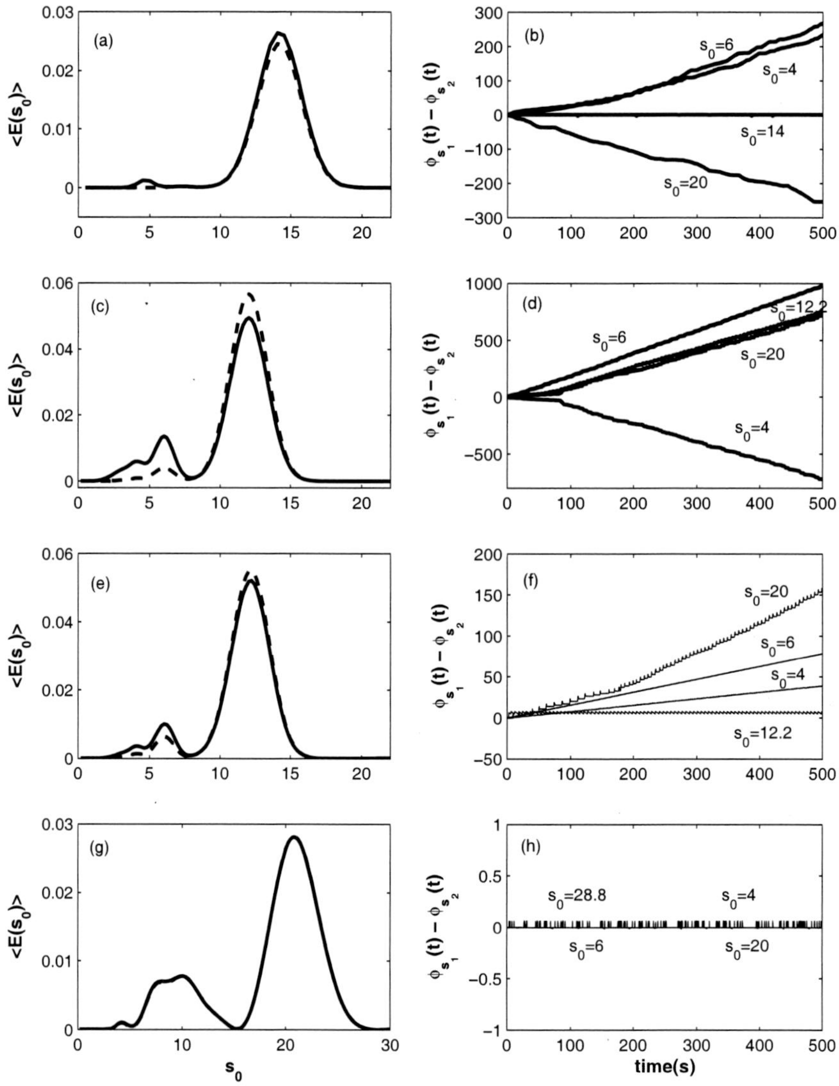

Figure 5. The normalized energy distribution in wavelet spectrum <E(s)> for the first (denoted by solid line) and second (by dashed line) model. The phase difference $\varphi s_1^{(t)} - \varphi s_2^{(t)}$ for the models. (a) Normalized energy spectrum of the wavelet transform for the modes I and II, (b) The phase difference between two models when they are couple unidirectionally, (c) Normalized energy spectrum at bidirectional coupling s'_1=0.1, (d) The phase difference at s'_1=0.1, (e) Normalized energy spectrum at s'_1=1.0, (f) The phase difference at s'_1=1.0, (g) Normalized energy spectrum at s'_1=6.5, (h) The phase difference at s'_1=6.5.

In this respect we have seen a new feature when the coupling is both ways. In such a situation the variation of <E(s_0)> against s_0 is plotted in figure (5c) when bidirectional coupling is s'_1=0.1. Here the coincidence of the normalized energy spectrum is not so evident as before but the plot of the difference {$\varphi s_1^{(t)}$ - $\varphi s_2^{(t)}$} does not vanish for any s_0 for any t. So that they are never scale synchronized. It should be noted that the two different picture presented above occur at two different value of coupling. In the bidirectionally coupled case a different situation is presented in figure (5d). When the parameter of coupling between chaotic systems increases, more and more time scales become correlated and one can say that the degree of the synchronization grows. For coupling strength s'_1=1.0, we observe that the normalized spectrum almost overlap again with very near peak at s_0=12.2 [in Figure (5e)], and so we observe that the phase difference remains zero for all time only for s_0=12.2 [in Figure (5f)]. With further coupling parameter increasing the regime of lag synchronization [LS] appears. Before the LS regime occurs the interesting phenomenon called as the intermittent lag synchronization may be observed. An arising of the lag synchronization between chaotic systems means that all time scales are correlated. In fact, from the condition of lag synchronization z(t-τ)≅w(t) one can obtained that W_z (s_0, t — τ) ≅ W_w (s_0, t) (and therefore $\varphi_{s_1}(t-\tau) \cong \varphi_{s_2}(t)$. It is clear from this that the phase locking condition | $\varphi s_1^{(t)}$ - $\varphi s_2^{(t)}$ | constant are hold. For large coupling strength s'_1=6.5, the two power spectra <E(s_0)> totally overlap [in Figure (5g)]. The time scale s_0=6, and s_0=20 which are not synchronized in the previous [Figure (5f)] are synchronized. The phase difference are remained very near to zero for all time and for all time scale s_0 where we have showed this variation for low s_0 and also for high s_0 value. It is clear to note that the difference will depend on the time lag τ. With further increase of coupling its leads to the decreasing of time lag τ and both the systems tend to complete synchronization i.e. z(t)≅ w(t) and corresponding phase difference $\varphi s_1^{(t)}$ - $\varphi s_2^{(t)}$ tends to zero for all time scales. It has been shown from above that the phase, lag, complete synchronization are mutually intercorrelated and different synchronization depends upon the number of time scales.

7. Measure of Synchronization for WA

The measure of synchronization ρ can be defined as the part of wavelet spectrum energy being fallen on the synchronized time scales

$$\rho_{1,2} = \frac{1}{E_{1,2}} \int_{s_{\min}}^{s_{\max}} \langle E_{1,2}(s) \rangle ds$$

where $[s_{\min}, s_{\max}]$ is the range of time scales for which the condition of phase locking are satisfied and $E_{1,2}$ is a full energy of the wavelet spectrum

$$E_{1,2} = \int_0^{\infty} \langle E_{1,2}(s) \rangle ds$$

For complete and lag synchronization this measure ρ is 1 and its value 0 for non-synchronized state. In the case of PS its value lies between 0 and 1 depending upon the synchronized time scales. The variation of ρ for different values of bidirectional coupling s'_1 are shown in figure (6).

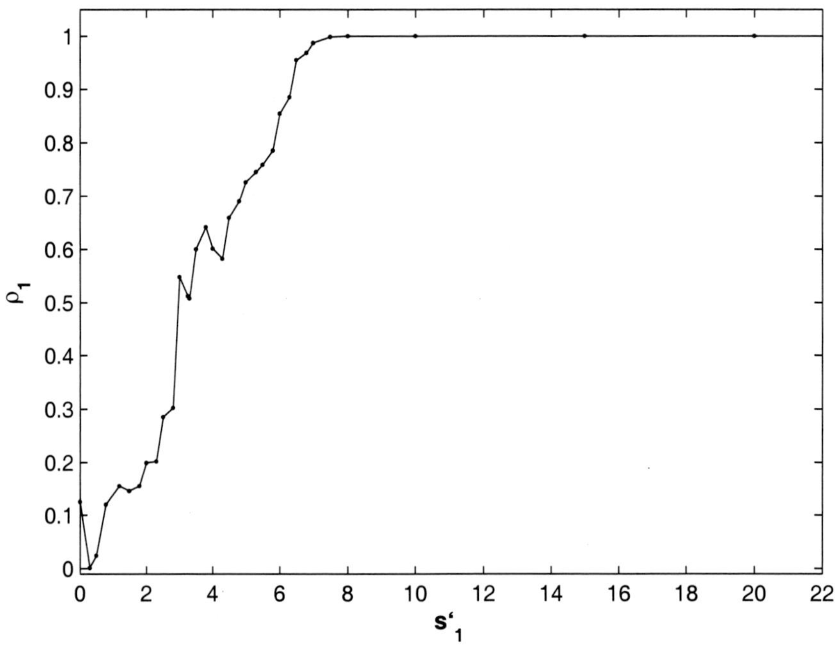

Figure 6. The dependence of the synchronization measure ρ_1 on different coupling strength s'_1 for the first model. The measure ρ_2 for the second model behaves in the similar manner, so it has not been shown in figure.

Summary and Conclusion

In this paper, we have studied the various type of chaos synchronization between two different nonlinear dynamical systems from the perspective of empirical mode decomposition and the wavelet theory. If the chaotic time series are characterize with the help of Fourier spectrum without main single frequency, the traditional techniques for detection of instantaneous phase are failed. For this chaotic time series data the corresponding phase space shows multiple center of rotations. It has been shown that phase plot with an unique center of rotation can be obtained from the intrinsic mode functions that are evaluated via the empirical mode decomposition and Hilbert transform. Before using the method of EMD, other approaches which have been successfully utilized by various researchers have been discussed and relevance of wavelet theory for the study of this type of synchronization is studied. It turns out that a new type of scale synchronization can be detected. We observe that where as the EMD method separates the complex signal in various IMF's corresponding to a definite frequency, the wavelet approach does the same but with respect to different scales. In particular the wavelet approach shows a different behavior for the unidirectional and bidirectional coupled oscillators. It is also observed that the state of synchronization strongly depends on the strength of coupling. Actual comparison is possible only through statistical methods [11] and is beyond the scope of the present communication. Incidentally, it may be mentioned that depending upon the strength of coupling a transition can take place from phase to complete synchronization. It is observed that wavelet theory predicts time scale synchronization due to the existence of scaling parameter s_0 and a similar behavior is shown by different intrinsic mode functions of empirical mode decomposition analysis that are actual modes for different frequencies.

Finally, it may be mentioned that different kinds of chaotic synchronization can be analyzed in either of the formalism of EMD or WA. According to our paper one can see that it will be a unified framework for time scale synchronization for any dynamical system. These two approaches are applicable for experimental data that is time series because it does not required priori information about the dynamical system. Such an analysis can prove highly useful in different branches of science such as electronics, medicine, signal communication, environment, biology, medicine and chemistry.

References

[1] A. Pikovsky , M. Rosenblum, and J. Kürths - Synchronization, A Universal concept in Nonlinear Systems, *Cambridge Nonlinear Science Series* 12 (Cambridge University Press, UK, 2001).

[2] M. G. Rosenblum, A. S. Pikovsky and J. Kürths - *Phys. Rev. Lett.*, 78 4193 (1997) .

[3] S. Boccaletti, J. Kürths, G. Osipov, D. L. Valladams and C. S. *Zhou* - *Phys. Rep.*, 366 1 (2002).

[4] N. E. Huang et al. - *Proc. R. Soc. London. Ser A* 454 , 903, (1998).

[5] D. Gabor- *J. IEE* London, 93 429 (1946) .

[6] M. Chavez, C. Adam, V. Navarro, S. Boccaletti, and J. Martinerie - *Chaos* 15 023904 (2005).

[7] R. Balocchi, D. Mnicucci, E. Santarcangelo, L. Sebastiani, A. Gemignani and M. Varanini - Chaos, Solitons and Fractals, 20 171 (2004), V. I. Ponomarenko, M. D. Prokhorov, A. B. Bespyatov, M. B. Bodrov and V. I. Gridnev - *Chaos, Solitons and Fractals*, 23 1429 (2005).

[8] G. Rilling, P. Flandrin and P. Goncalves- In *Proc. IEEE-EURASIP workshop on Nonlinear Signal and Image Processing* NSIP-03 (2003) .

[9] A. A. Koronovskii and A. E. Hramov- JETP Letters 79(7) 316 (2004), *J. Comm. Tech. and Electronics*, 50(8) 894 (2005).

[10] A. A. Koronovskii, M. K. Kurovskaya and A. E. Hramov- Technical Physics Letters, 31(10) 847 (2005), *JETP*, 100(4) 784 (2005).

[11] M. L. Van Quyen, J. Foucher, J-P. Lachaux, E. Rodriguez, A. Lutz, J. Martinerie and F. J. Varela- *Journal of Neuroscience Methods*, 111 83 (2001).

In: Progress in Chaos Complexity Research
Editors: Franco F. Orsucci and Nicoletta Sala
ISBN: 978-1-60456-375-7

Chapter 8

RECURRENCE QUANTIFICATION ANALYSIS, VARIABILITY ANALYSIS, AND FRACTAL DIMENSION ESTIMATION IN ^{99M}TC-HDP NUCLEAR SCINTIGRAPHY OF MAXILLARY BONES IN SUBJECTS WITH OSTEOPOROSIS

***Elio Conte*[1], *Giampietro Farronato*[2], *Davide Farronato*[3], *Claudia Maggipinto*[4], *Giovanni Maggipinto*[5] and *Joseph. P. Zbilut*[6]**

[1]Department of Pharmacology and Human Physiology, University of Bari (Italy), and TIRES - Centre for Innovative Technologies for Signal Detection and Processing, and ICS – International School for Studies on Radioactivity, Nuclear and Non Linear Methodologies

[2]Department of Orthodontics, University of Milan, Via Commenda 10, Milan, Italy

[3]Dentistry, DDS, PHD, Associate in Oral Surgery, University of Milan, Via Commenda 10, Milan, Italy

[4]Dentistry, DDS, Resident in Oral Surgery, University of Milan, Via Commenda 10, Milan, Italy

[5]Nuclear Medicine Service – Di Venere Hospital – Carbonara-Bari (Italy)

[6]Department of Molecular Biophysics and Physiology, Rush University, Chicago, IL 60612,USA

ABSTRACT

We develop a non linear methodology to analyze nuclear medicine images. It holds on Recurrence Quantification Analysis (RQA), on Analysis of Variability by variogram, and on estimation of Fractal Dimension. It is applied to five subjects with osteoporosis in comparison with five control subjects. Bone nuclear images are obtained after administration of ^{99m}Tc-HDP. Regions of interest (ROI) are selected in maxillary bones of oral cavity. Some basic non linear indices are obtained as result of the methodology and they enable quantitative estimations at the micro-structural and micro-architectural level of bone matrix investigated. The indices result very satisfactory in discriminating controls from subjects with osteoporosis. Their appear of interest also in the case of dentist intervention often engaged to evaluate oral signs and, in particular, to utilize mandibular or maxillary bones indices in relation to possible loss of bone mineral density and/or to microarchitectural deterioration of bone tissue.

1. INTRODUCTION

As it is known, X-ray images represent the most conventional radiograph tool used from dentists for diagnosis affecting teeth and bones of oral cavity. In fact, radiographic investigations are able to display with accuracy differences in bone density in a given anatomic region and to provide images of satisfactory high resolution. However, as correctly outlined by various authors [see in particular, R.I. Ferreira et al. Ref.1], conventional radiological techniques evidence morphological changes. They essentially detect bone changes within 30-50 percent bone mineral content. Consequently, the determination of the extent of some bone lesions by radiological tools may result in some cases rather problematic or even erroneous.

Nuclear bone scintigraphy is obtained by radiopharmaceuticals. It is able to evidence reactive modifications in osteoblastic activity that may remain hidden in radiographic images. It does not evidence with optimization the morphologic changes and, generally speaking, its resolution is not so high as in radiograph images, but nuclear bone scinitigraphy results to be positive if there is a 10 percent increase in osteoblastic activity respect to the normal case. Consequently, it represents a decisive technique when one has to analyze the dynamic profiles of the metabolic process under consideration.

As usually in nuclear medicine, bone scintigraphic images are obtained following intravenous administration of a radiopharmaceutical obtained by addition of the radioactive ^{99m}Tc to diphosphonates. Technetium -^{99m}Tc is a radionuclide that has a very short physical half life of about six hours, and its gamma energy is about 140 KeV. Therefore, the administered dose of radionuclide-labelled tracer, consisting in about 600 MBq, does not involve severe problems of radioprotection. The kinetics of such radiopharmaceutical is well known: it circulates through the blood stream until it is fixed mainly into the sites of active bone turnover and soft tissues. Finally, it is eliminated in the urines[2].

Starting with 1981, some of us, E.C. et al., [see in detail ref. 3], were able to realize the kinetic compartment model for ^{99m}Tc-phosphonates, in particular Technetium-99m DPD and Tc—99m (Sn) MDP. The use of such compartment model, fitted to a number of experimental data, resulted particularly advantageous to calculate in detail all the basic parameters involved in the kinetics of such two ^{99m}Tc- phosphonates fixing the basic dynamical features characterizing the mechanism of fixation in vivo of such ^{99m}Tc- phosphonates marked in bone. Actually, their final understanding of the mechanism of fixation in bone, remains still incomplete but the main feature, largely accepted, is that such radiopharmaceuticals are absorbed to the amorphous bone.

Consequently, there is a feature of bone scintigraphy, performed by such compounds, that remains very important: the mentioned radioactive uptake remains strongly correlated with the rate of bone mineralization. Therefore, bone scintigraphy with such radiopharmaceuticals should be considered more a dynamic rather a static investigation under the profile of the general physiology.

In conclusion, from one hand we have in our studies radiographs plan images, tomography, magnetic resonance imaging that must be considered essentially structural imaging techniques. On the other hand, we have bone scintigraphy that remains a functional method. It represents a valid technique to analyze early physiologic changes given as result of biochemical alterations. By such nuclear method, they may be detected also before significant bone mineral changes are evidenced by other methodologies.

In these conditions, the reason of the present study becomes clear in itself. It is known that many oral diseases evidence metabolic changes in the oromaxillofacial complex.. Therefore, the nuclear bone scintigraphy may have some direct and important indications. In order to support such conclusion, the aim of this study was to evaluate the possible relationship between osteoporosis and oral signs given by nuclear scintigraphy of maxillary bone in a group of subjects affected by such disease.

2. The Methodology

We must give the first look at the methodologies that will be employed in the present paper.

In vitro studies have evidenced that [1] the competitive adsorption of ^{99m}Tc labelled diphosphonates to pure hydroxyapatite is forty times greater than to pure organic bone matrix.

In other terms the ^{99m}Tc labelled diphosphonates is a radioactive process of uptake that correlates very well with the rate of bone mineralization. Here there are involved processes that, under the dynamic profile, result highly non linear and non stationary. In particular, the osteoporosis is an abnormal physiological condition characterized by a loss of bone mineral density and micro alteration and deterioration of bone tissue.

The complex of such features indicates that the way to analyze adequately the maxillary bone images obtained by ^{99m}Tc-labelled by diphosponates, requires to use the non linear methods of analysis.

In this paper we will use substantially three methods:

1. The variogram technique that we call Variability Analysis;
2. The Recurrence Quantification Analysis that in literature is abbreviated usually by RQA;
3. The Fractal Dimension Estimation.

Let us consider the variogram. This is a substantially novel methodology that some of us (E.C., and J.P. Z) introduced recently to analyze the variability of series of data in space or in time domain [4]. For a review of the method and for previous applications of variogram in various fields see in detail the ref.[4].

The second method is that one of the Recurrence Quantification Analysis (RQA). It is largely known in literature for analysis of non line series of data given in space or in time. Starting with 1994, it was introduced by some of us, J.P.Z et al [5]. A lot of RQA applications have been given in literature and a satisfactory explanation of the basic principles and methods of RQA may be found as example in Ref.[6] and references therein.

Finally, we calculate the fractal dimension of the analyzed nuclear bone images. We will not use here the method of the variogram as it was explained and used in [4] but we will utilize directly the estimation by Hurst exponent in order to speed up our applications.

These are essentially the three methods that we will use in the present paper.

Let us start giving some detail on the analysis of variability.

The greatest part of physical, biological and physiological phenomena in nature exhibit a non linear and a non stationary behaviour. Very often the presence of a non linear dynamics in a process is indicative of what is usually called a very complex dynamical regime. Such processes are often identified to be divergent or chaotic.

As a guide that may aid in explaining what we mean by variability, let us consider some examples of time series that pertain to our case. If we collect the data in time of a recorded ECG signal or of the R-R intervals, starting with its QRS complexes, or of an EEG, the behaviour of such data in time evidence fluctuations. The data exhibit irregular, complex, and seemingly random dynamics. The variability of data relates thus such irregular dynamics that falls under our consideration.

The variogram, see [4], is the tool that we may use to analyze the variability of the data in spatial or time domains.

In practice, the variogram is estimated by

$$\gamma(h) = \frac{1}{2n(h)} \sum_{i=1}^{n(h)} \left[x(u_i) - x(u_i + h) \right]^2 \tag{1}$$

where $n(h)$ is the number of pairs of series data at lag distance h (in space or in time), and $x(u_i)$ and $x(u_i + h)$ are the series of data located at space or time u_i $(i = 1,2,...)$ or at increments $u_i + h$ $(h = 1.2,......)$.

We must now establish how to apply such method to the sphere of nuclear medicine. We will expose a method that obviously will may be applied for the elaboration of a variety of nuclear radiopharmaceutical scintigraphies.

Let us consider the particular case of our bone scintigraphic image. Let us consider to have selected a bone Region of Interest, a R.O.I, that obviously will be reduced to a matrix of dimension ($m \times n$). In our case we selected a bone ROI of 186×73 that consequently gives a total of 13578 points-pixels for the considered bone matrix. This choice was suggested only from reasons of convenience in this our preliminary study but we could have considered also different dimensions of the ROI in respect of any other experimental requirement. In brief we explored bone regions of 23.63 mm $\times$ 9.77 mm with an unitary section of 0.016 mm^2.

Note that in the case of bone scintigraphy we are not concerned with time series, that is to say with biosignals as ECG or EEG recorded in time. In the case of a bone selected ROI we have an image composed by coloured pixels located in a bidimensional space. The coloured image in the ROI will correspond proportionally to the radioactive activity fixed in the biological matrix taken in consideration and represented by the image in the ROI. In detail, by our bone ROI, we are concerned with the fluctuations or inhomogeneous distribution of radioactive activity corresponding to the radiopharmaceutical uptake in the different located bone sites during the uptake process of the ^{99m}Tc-labeled diphosphonates. The reason of our study is to estimate in detail the variations of such radioactive distribution, site by site, within the selected bone ROI. In particular there will be sites that, also at different space locations, could show very similar bone uptakes and sites, having different locations, that instead will show evident variability, or, that is to say, a significant fluctuation in the measured values of radioactive activity. Let us remember in fact that ^{99m}Tc-labeled diphosphonates correlates well with the rate of bone mineralization. Consequently, in a selected bone ROI, we expect to find space variations or, equivalently, space fluctuations in reason of the physiological conditions of the explored bone matrix and, in addition, in reason of the altered conditions of radiopharmaceutical uptake in function of the pathology under examination.

Let us make now a further step on. In order to examine the image represented in the selected ROI we need now to have a series of data in the bidimensional space domain. In detail we will have a three columns set of data in which the pair of data in the first two columns will give the (x,y) location of the pixel under consideration in the fixed ROI (representing the bone matrix in consideration) while the remaining data in the third column will represent the values of a variable that characterizes the coloured image that we have in examination in the selected ROI [for details see still ref.4]. In order to calculate the values of variables characterizing the coloured image, we may utilize a MatLab R12 software. By it the coloured bone image, selected in the R.O.I, will be reduced to the three fundamental colours as it happens for each image in nature. It is known in fact that they are three basic colours in nature, the Green, the Red, and the Bleu, respectively. In conclusion our image will be resolved in three series of 13578 points in space domain and each point will represent respectively the Green, or Red or Bleu intensity linked to its (x,y) space location in the selected bone ROI (for details see still ref.[4]). Two subsequent colour intensities in the series (for Green, Red, and Bleu) will have lag distance $h = 1$ and so on progressively for the other remaining values. We will use the following expression for lag distance

$$h_k : \ k = 186(n-1) + n \times m \tag{2}$$

In this manner, finally, we will be re conducted to three space lag-series

(*h Green intensity*);
(*h Red intensity*);
(*h Bleu intensity*);

by which we will be now able to apply our variability analysis pixel by pixel using the (1).

By using the (1) we will analyze the variability or similarity of ^{99m}Tc-labeled diphosphonates bone uptake for distant pixels with distances varying in function of the selected space lag h. In our study we selected the space lag h to vary from 1 to 2716 and thus examining simultaneously bone uptakes for adjacent pixels as well as for increasing distances until 2716 standard space lag unities. In this manner we will have the behaviour of the Technetium labelled diphosphonate distribution in the inner of the matrix bone that we selected as ROI. Similar values of the distribution of the radiopharmaceutical will give low values of the variogram and this will mean that in such space delayed regions of the bone matrix we will have a rather uniform distribution of accumulated radioactivity and thus of radiopharmaceutical. In brief, in such regions we will have what we may call recurrent values of bone uptake or a rather homogeneous distribution of the radiopharmaceutical. Instead, space separated regions of the ROI in which the variogram will exhibit relatively high values will correspond to sites of bone matrix showing a profound variability and thus a large fluctuation of the distribution of the radioactive activity and thus of the uptaken radiopharmaceutical.

Since, as previously discussed, nuclear images of bone, performed by Technetium labelled diphosphonates, are functional images linked to physiological changes that are direct consequence of biochemical alterations, being the nature of such processes highly non linear and non stationary, we expect to find a great variability in bone districts whose functional regime is normal and we expect to find instead a decreased variability in condition of pathology.

This is briefly the conceptual scheme one must expect by performing analysis of variability for processes characterized from high non linearity.

We may now pass to expose the method of RQA. For brevity, we will not re-enter in the theoretical developments here (for details see ref.[5]]. We will recall only some brief conceptual foundations.

It is mainly conceived for analysis of non linear and non stationary series embedded in phase space with data given in time or in space.

A unique feature of series with seemingly irregular dynamics is the quite general absence or reduction of periodicity. In order to develop analysis in such conditions, starting with 1987, Eckmann et al. [7] introduced the idea that significant periodic structures might be uncovered in physical processes by mathematically embedding the ordered series in higher dimensional space and then realizing a criterion for determining what must be considered a recurrence of a point in the given series. These points can be plotted on a symmetric matrix and, as example, deterministic and thus non random regimes may be observed in the plot by short line segments parallel to the main diagonal. The embedding procedure for phase space reconstruction is then the key feature in the methodology. The new graphical tool introduced by Eckmann et all [7] was called a recurrence plot (RP), and it is based on the computation of the distance matrix between the reconstructed points in the phase space. Once a distant matrix is calculated it may be displayed by darkening the pixel corresponding at specific coordinates corresponding to a distance value lower than a predetermined cutoff, usually identified as a ball with a prefixed centre and radius. The darkened points individuate recurrences of the dynamical process under investigation and the recurrent plot provides insight into periodic structures and clustering that obviously escape to the direct observation of the given series of data. In order to extend the procedure and render it more quantitative Webber and Zbilut [5] developed the methodology called RQA. As result, they introduced several variables to quantify RP. Since they will be used in the present paper it becomes important to add a brief comment of each of such several variables here.

The first variable is % Rec that states for % Recurrences. It defines the percentage of darkened pixels in the recurrence plot.

The second variable is the % Det that states for % Determinism and expresses the percentage of recurrent points forming diagonal lines.

The third variable is the Entropy representing a Shannon entropy of line segments distributions.

The Fourth variable is the inverse of Max Line representing the reciprocal of the longest diagonal line segment and it will relate directly to largest positive Lyapunov exponent and thus it will characterize the divergence of the system to which the series of data is connected and thus, in some sense, its chaoticity.

The mentioned variables will quantify the detereministic structure and complexity of the series of data and thus of the corresponding process under consideration. In particular, % Rec will quantify the amount of cyclic behaviour. % Det will characterize the amount of determinism, the Entropy will express the richness of deterministic structuring. Other variables were also considered in [5] but we will not take in consideration them here. Other variables as % Lam. that states for % Laminarity, T.T. that states for Trapping Time were introduced by N. Marwan [5]. Also the Ratio is often considered and it is given as ratio of % Det. with % Rec.

The final stage of our methodology will consist in the estimation of the fractal dimension of bone uptake of ^{99m}Tc labelled diphosphonates in the considered ROI. It is well known that the fractal structure of bone has been previously outlined by various authors [8]. Our evaluation in the present paper could have used the procedure that was exposed previously in ref.4 but we will use directly the Hurst exponent for brevity. Our estimation will exhibit some features of novelty since, as it results to our knowledge, estimations of fractal dimensions of radiopharmaceutical relating bone uptakes were not previously obtained from other authors in bone ROI as well as in normal conditions and for pathology.

3. Experimental Methods

The study was based on ten post-menopausal women varying in age from 43 to 73 years to years (five subjects with osteoporosis and five controls). The subjects were all in natural menopause. Each subject had not bone metabolism diseases as hyperparathyroidism, hypoparathyroidism, Paget's disease, osteomalacia, renal osteodystrophy or imperfect osteogenesis or cancer with bone metastasis or still significant renal diseases. They were not assuming specific drugs or hormones. All subjects were previously analyzed and classified by using BMD. All the data regarding menopausal status, age, weight and height were previously collected by compilation of an ad hoc questionnaire. All the patients did not have menstruation from one year at least.

A calibrated tracer dose of 600 MBq of ^{99m}Tc-HDP was prepared and intravenously injected to each subject. Bone scintigraphy was performed about 3 hr after injection using a nuclear gamma camera Siemens (E. Cam. double head). The obtained bone images scintigraphies were than elaborated according to the methodology exposed in section 2.

4. Results

As an indication, in Figures 1 and 2 we give the results of cranial bone images as they were obtained for a control subject and an osteoporosis patient, respectively. Similar scintigraphic figures were obtained for the remaining subjects.

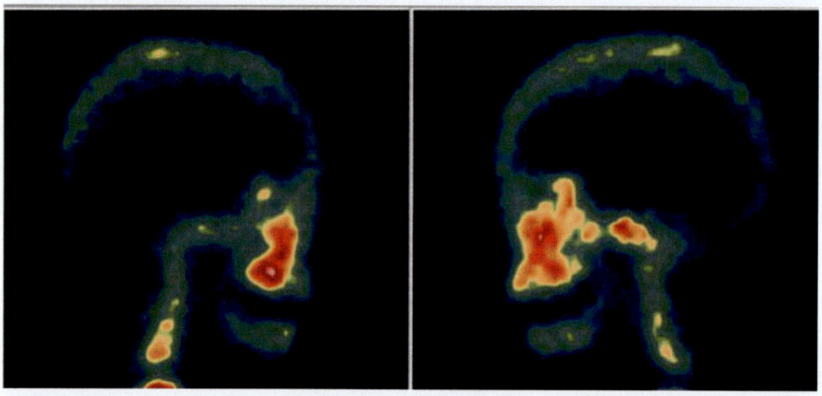

Figure 1. Subject Control: Sch. Bone scintigrafy. Nuclear image display obtained by ^{99m}Tc-HDP.

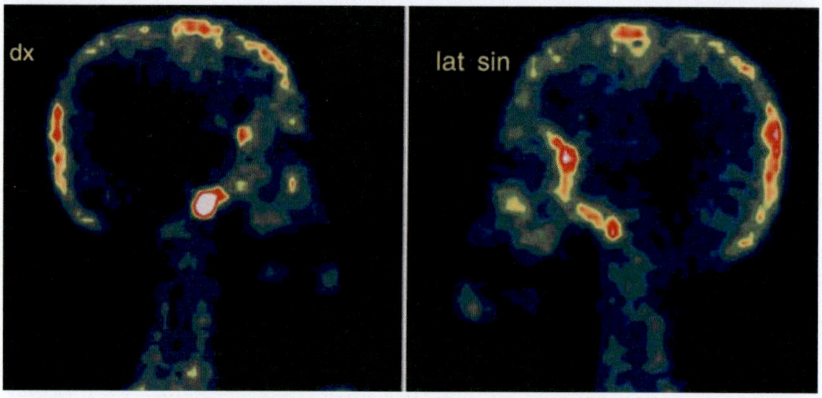

Figure 2. Subject with Osteoporosis: Dem. Bone scintigrafy. Nuclear image display obtained by ^{99m}Tc-HDP.

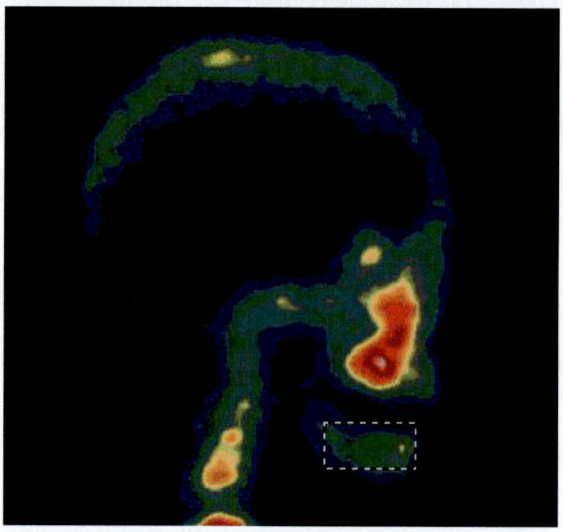

Figure 3. Subject Control: Sch. Bone scintigrafy. Nuclear image display obtained by ^{99m}Tc-HDP. Example of selected R.O.I.

In Figure 3 we give, as example, the result of a selected mandibular region of interest. A similar procedure was followed for all the subjects of the investigation and for all the considered maxillary bones. Four regions were considered respectively: IL that states for Lower Left, IR that states for Lower Right, SL that states for Upper left and SR that states for Upper Right.

As previously explained in the section 2, all the obtained images in the bone selected ROI were than analyzed by the MatLab R12 Software. In this manner the series relating the three basic colours, Green-Red-Bleu, and their space location were calculated. Some examples of results are given in Figures 4, 5, 6, 7.

Figure 4 gives Green intensity of the selected ROI (IR) for a normal subject. As previously said, it contains 13758 points having a definite space location to which a lag space may be connected by the e (2). Green intensity, combined with the corresponding Red and Bleu intensities for each (x_n, y_n) point, results to be proportional to the measured radioactive activity of ^{99m}Tc-labeled diphosphonates as it has been accumulated and distributed in the different sites in the considered bone matrix. Figure 5 gives Green intensity for the first 3381 points series. This fragmentation has been introduced to enable the reader to inspect in detail the complex behaviour of such colour to which corresponds, in combination with the behaviour of the other colours, the spatial distribution of the radioactivity for the considered subject. Figure 6 represents instead the Red intensity for a subject with osteoporosis in the same selected ROI. Also in this case Figure 7 gives a partial behaviour of spatial distribution corresponding to only 3150 points. All such figures are given here as examples. A simultaneous inspection of Green, Red and Blue colours gives immediately the possibility to ascertain bone uptake differences of the radiopharmaceutical in control respect to subjects with osteoporosis.

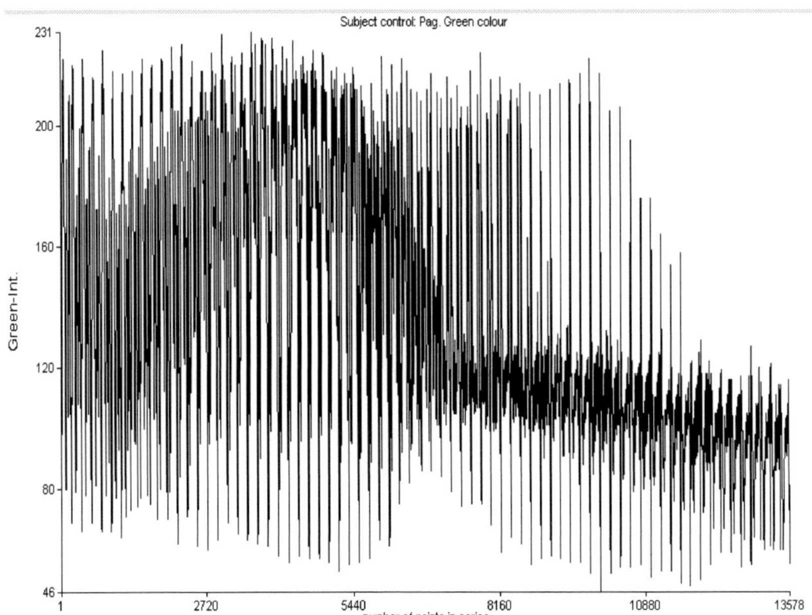

Figure 4. Behaviour of space series for green colour in normal subject with 13578 points. Similar series may be obtained for red and bleu colours respectively.

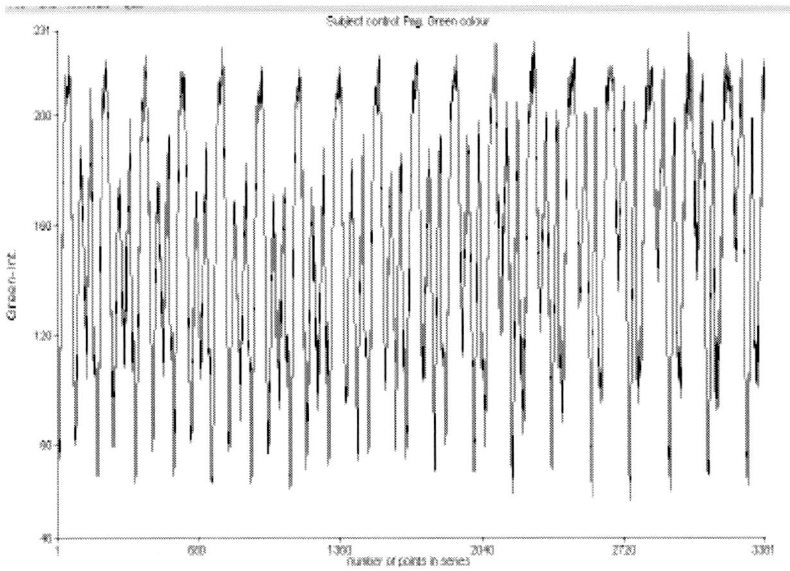

Figure 5. This is the same previous series (Figure 4) with 3381 points.

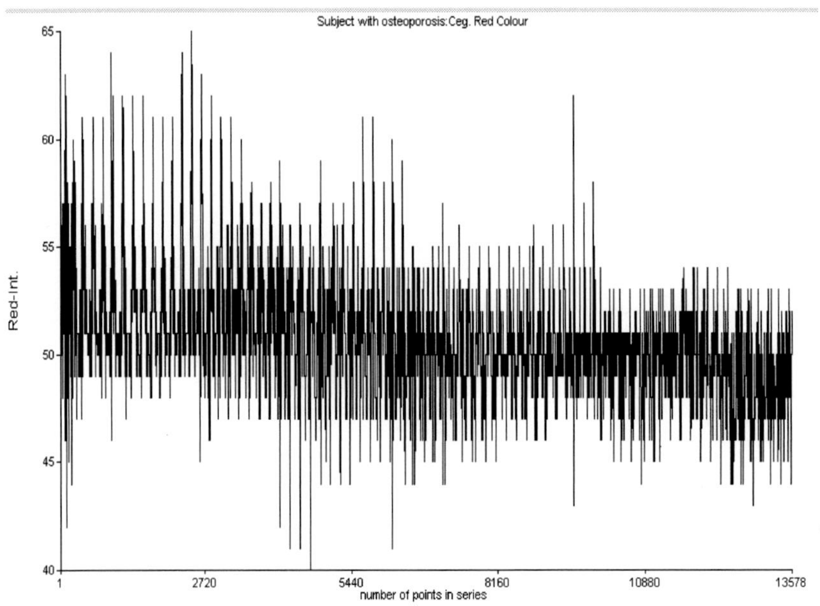

Figure 6. Behaviour of space series for red colour in subject with osteoporosis with 13578 points. Similar series may be obtained for green and blue colours.respectively.

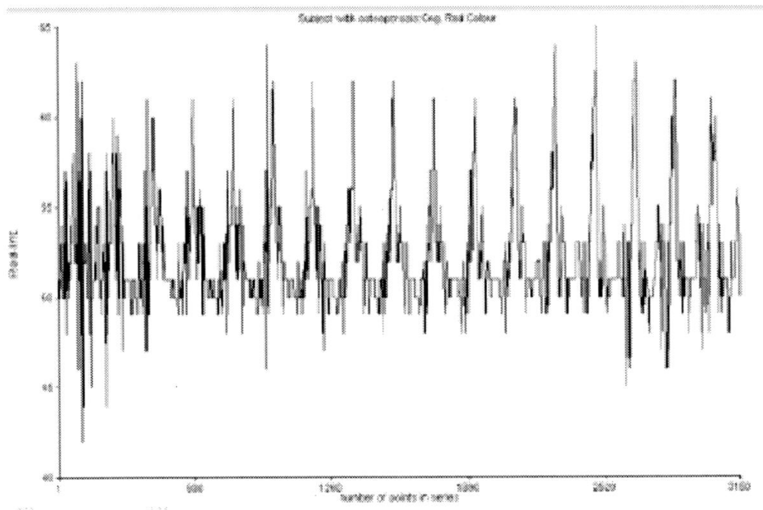

Figure 7. This is the same previous series (Figure 6) with 3150 points.

We may now pass to examine the results on variability analysis obtained by variograms. They were calculated by using the (1), and the (2). As usually, we give some examples of the obtained results. In Figure 8 we have the behaviour of the variograms in the case of controls and subjects with osteoporosis, respectively. The results are given for IR Bleu, Red and Green. Variograms were realized by using normalized data. We normalized the series by subtracting the series mean and dividing by the series standard deviation. All the variograms were calculated for spatial lags ranging from 1 to 2716. Remember that each lag value represents the distance between different bone sites of ^{99m}Tc-labeled diphosphonate uptake. Note that a strictly periodic series shows a periodic variogram. In a more complex regime we have a wave variogram with oscillations of decreasing amplitude around a plateau that often is called the sill. The oscillation reflects periodicity or, more correctly, recurrences in the data of given distance fixed by h. They correspond visually to the lower values assumed by the variogram. The upper values of the variogram delineate instead the regimes of greatest variability in the data. Figure 8 shows that we have distant sites in the bone matrix that exhibit recurrences and this is to say that their behaviour respect to the uptake process of ^{99m}Tc-labeled diphosphonate results very similar. As well as we have also great regimes of variability. This spatial dynamics occurs in the case of controls as well as in the case of subjects with osteoporosis at the varying values of the spatial lag h. We have bone sites that at a certain distance exhibit recurrent values in the uptake of the ^{99m}Tc-radiopharmaceutical while they exhibit great variability at other certain distances. By comparison, in Figure 9, it is seen that we have greater variability and lower recurrence in the case of normal subjects and lower variability and greater recurrence in the case of subjects with osteoporosis. Let us remember that by ^{99m}Tc-labeled diphosphonate uptake we study the correlation of uptake with the rate of bone mineralization. In addition, osteoporosis is a pathological condition characterized in primis by a loss in bone mineral density and by a micro-architectural deterioration. As it is easily verified, the results obtained by variability analysis, strongly evidence and quantify all such processes that are in course in the case of controls and of subjects with osteoporosis. Once again the use of variability analysis confirms its power in prediction and interpretation of physiologic data.

Controls

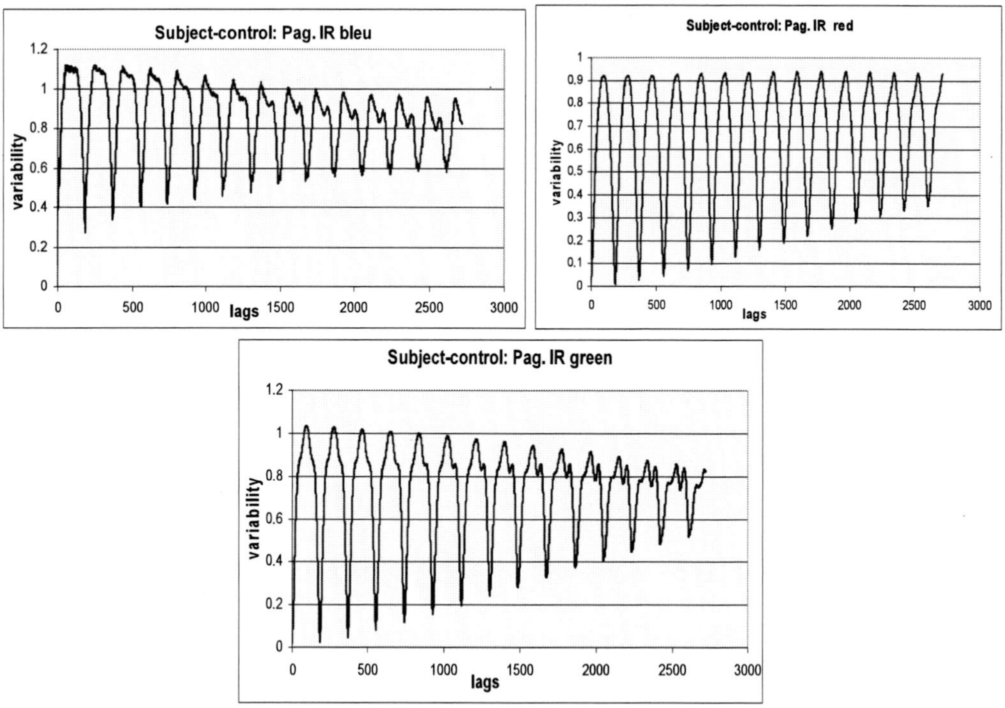

Subject with Osteoporosis

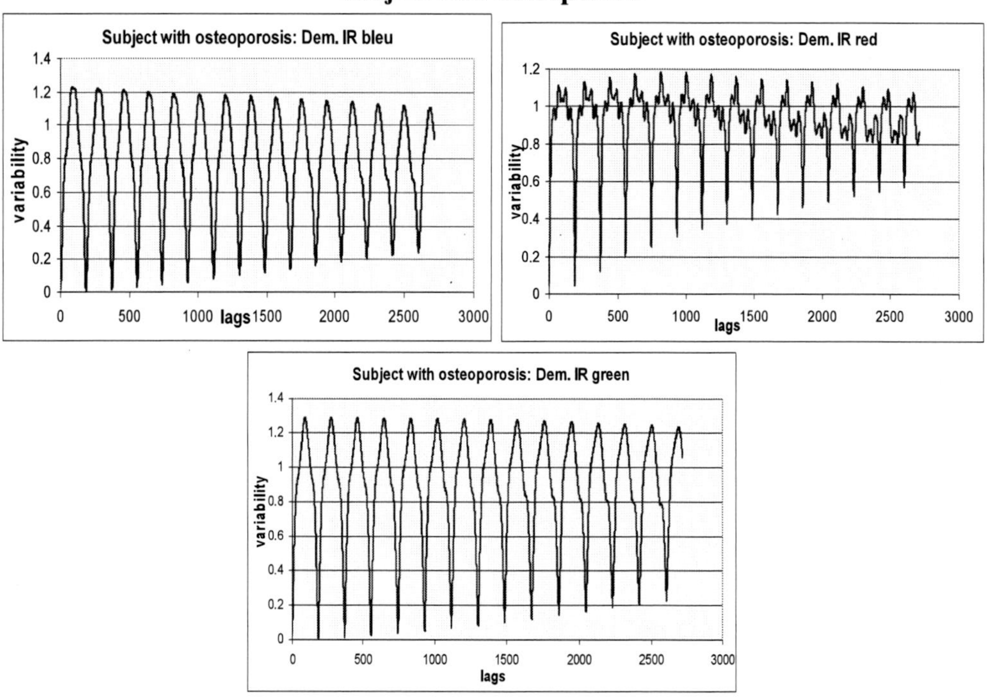

Figure 8.

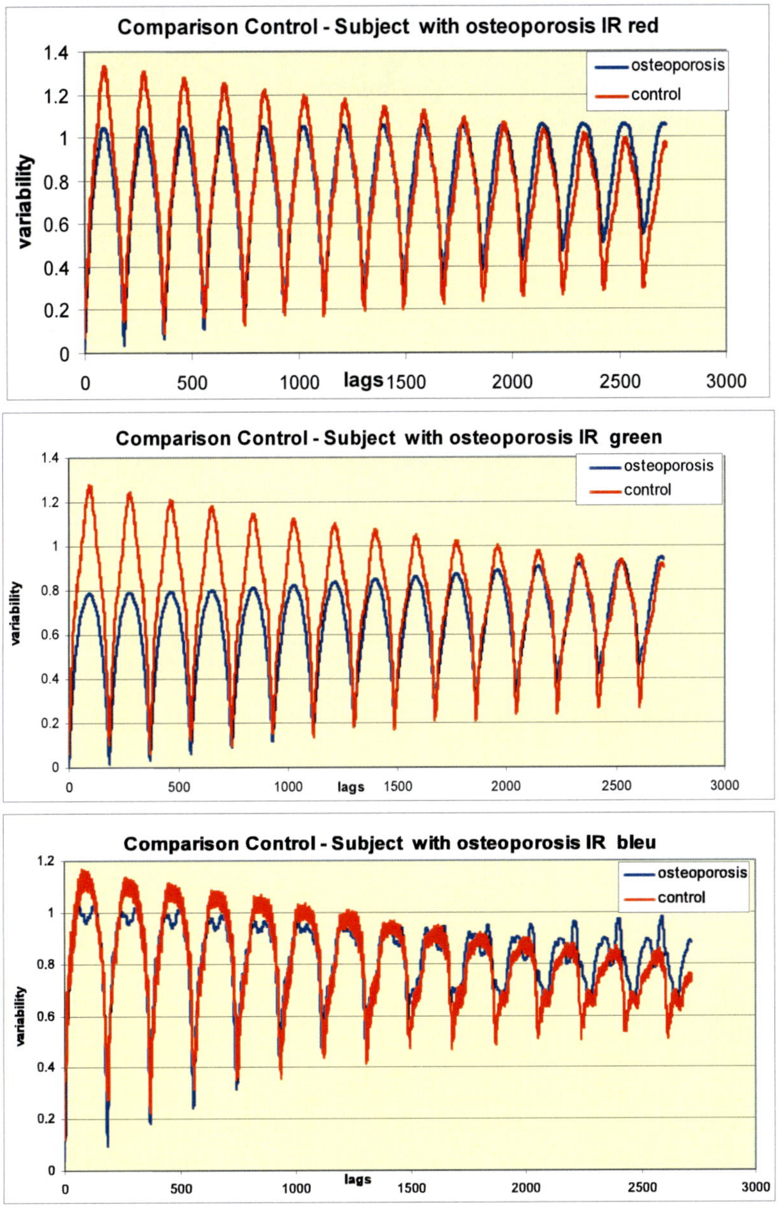

Figure 9.

There is still another evidence that deserves to be outlined here. When one uses RQA for analysis of time or space series, may introduce also RQI that is the quantification of percent Recurrent Intervals in the given series. It results very similar to our variability analysis also if with some basic differences. Generally speaking, RQI behaviour results very similar to variogram with the only difference that, in the plot, one is obviously reversed respect to the other.

In figures 10, 11, 12 we calculated RQI for normal and subjects with osteoporosis and, for brevity, we limited our analysis to only 450 lags. We did not use in this case normalized data. From the comparison of RQI with the previous variograms, the equivalence of the two methods is evidenced and, in particular, the use of RQI enables to identify in detail the basic differences that occur in the uptake mechanism of ^{99m}Tc-radiopharmaceutical in bone in the case of normal subjects respect to those with osteoporosis.

At this stage of our analysis, we should now complete our investigation on variability introducing some proper variogram indices to quantify the differences that we observed between bone uptake of controls and subjects with osteoporosis. For brevity, we choose to defer such our analysis directly to the following paper that is in progress, and, overcoming this stage, we pass to introduce directly the results that we obtained by using RQA.

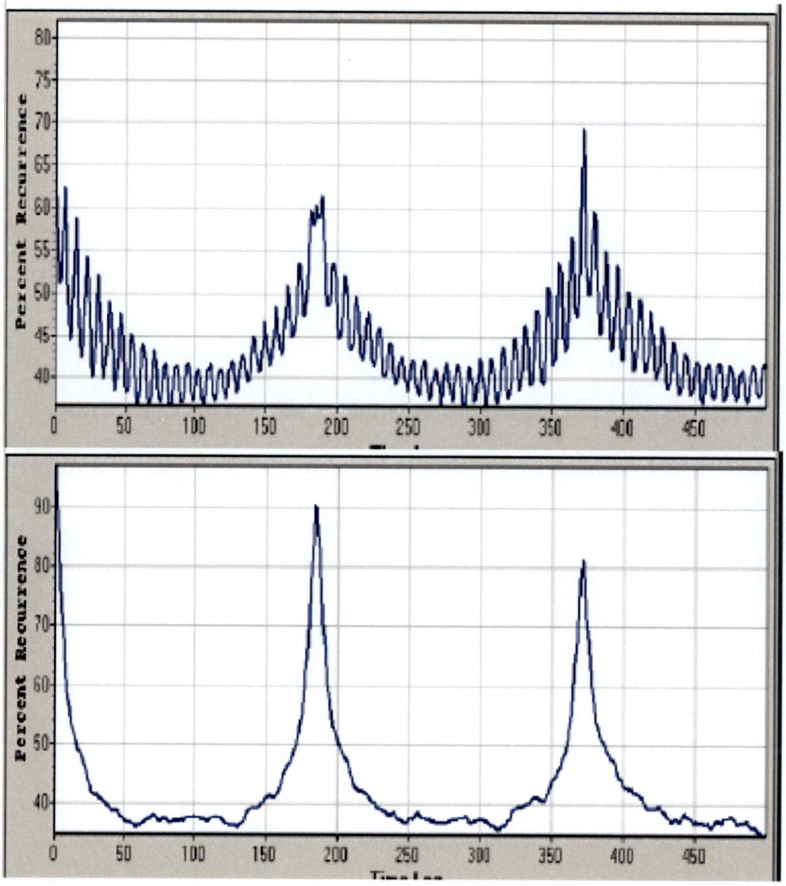

Figure 10. RQI of control (Cap.) and subject with osteoporosis (Mas.) with space-lags ranging from 1 to 500. Bleu colour. The equivalence of RQI results with previous variability analysis is evidenced.

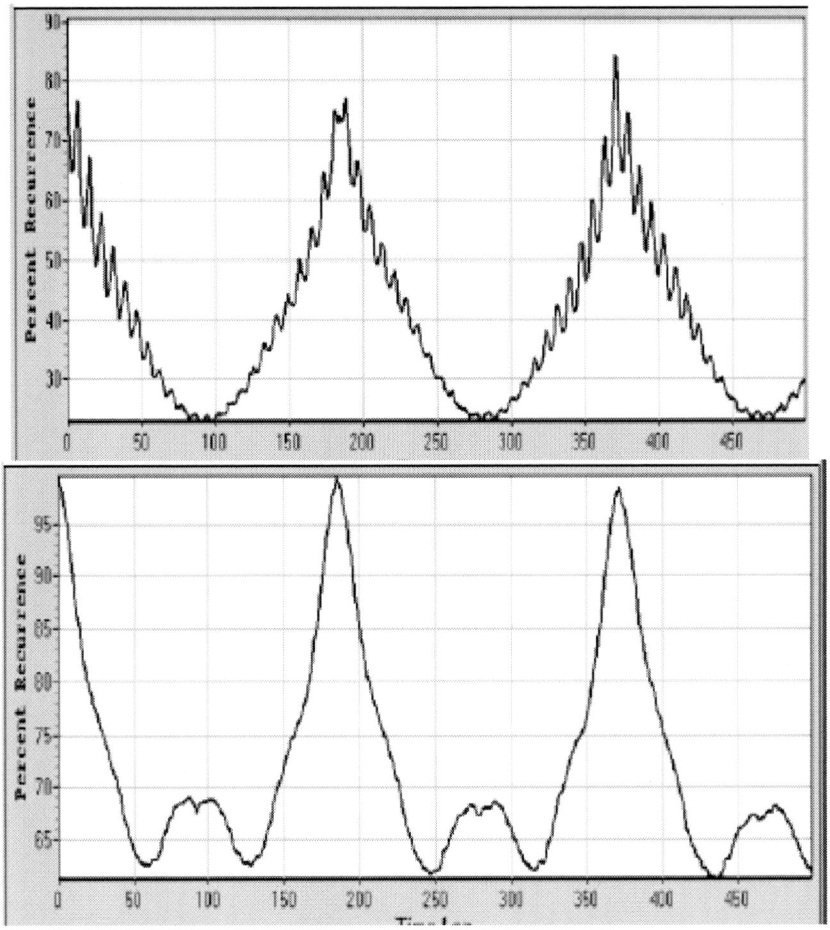

Figure 11. RQI of control (Cap.) and subject with osteoporosis (Mas.) with space-lags ranging from 1 to 500. Red colour. The equivalence of RQI results with previous variability analysis is evidenced.

Such non linear analysis was performed on the three basic series, Green-Red-Blue starting our preliminary study with the following parameters: radius was selected R=1.5, the line length L was chosen L=2, delay and embedding dimension were taken 1. The results are given in Tab.1 for normal subjects and in Tab.2 for subjects having osteoporosis. Figures 13, 14, 15, 16 give comparison of controls and subjects with osteoporosis in all the cases under our experimental verification.

Let us give some results. For IR, % Rec gave 1.708 ± 0.685 for Red, 1.515 ± 0.527 for Green and 3.214 ± 0.408 for bleu for the controls. In the case of osteoporosis we had 6.452 ± 4.392 for Red, 3.599 ± 2.727 for Green and 2.426 ± 0.607 for bleu. Significant statistical differences resulted with p=0.04. Similar results were obtained for IL, SR and SL.

Very significant differences were obtained for % Det. for IR, IL, SR, SL. In the case of normal subjects it ranged from about 11% to 15%. In the case of subjects with osteoporis it ranged from 39% to 50%.Obviously the results must be compared also ROI by ROI, that is to say, as example, IR for normal subjects respect to osteoporosis.

Significant statistical differences resulted with p=0.02.

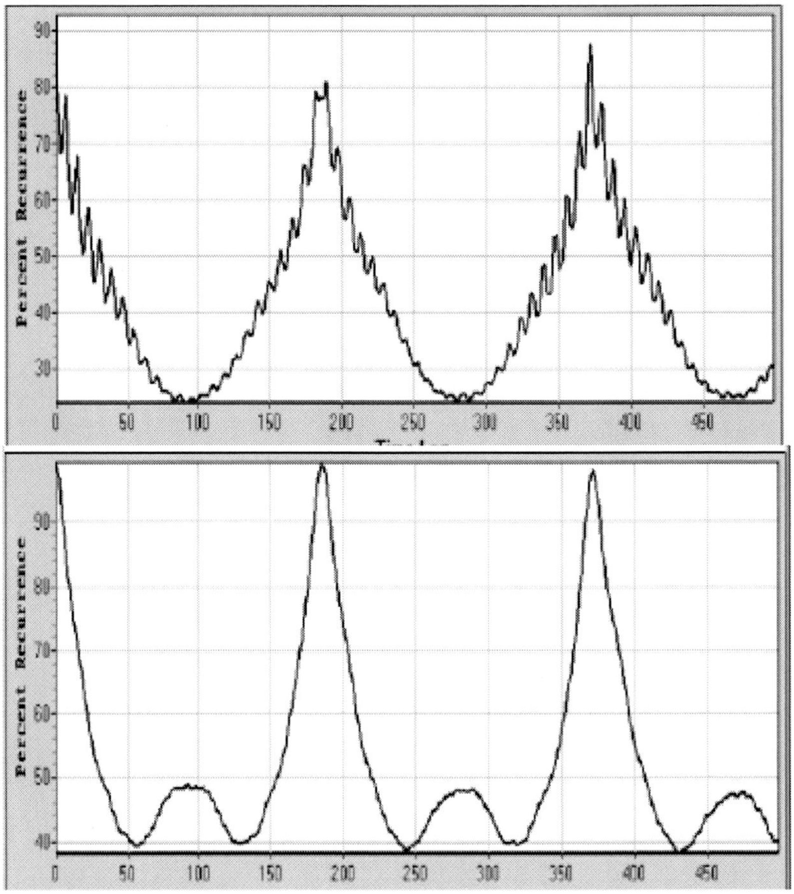

Figure 12. RQI of control (Cap.) and subject with osteoporosis (Mas.) with space-lags ranging from 1 to 500. Green colour. The equivalence of RQI results with previous variability analysis is evidenced.

Another important result was obtained for %Lam. We did not define it previously just to evidence it now before the exposition of the results that we obtained. Usually, for time series % Lam and T.T represent fast transitions or sudden changes from one state to another. % Lam represent instabilities and T.T is the actual time spent in the transition. This holds for time series but the same applies to series in space domain. In spatial dimension, we only speak of spatial respect to time transition. In particular it remains the concept of instabilities that in our case results very important. In fact, in our case strong instability should indicate that the bone site uptake by ^{99m}Tc-labeled diphosphonate has become so instable respect to uptake and maintenance of the radiopharmaceutical owing to the presence of a profound alteration or deterioration of the same feature in bone tissue in the course of the pathology. Inspection of Tables 1 and 2 indicates that in the cases of controls we had %Lam values ranging about 17-22% while instead it increased to 47-58% in the case of osteoporosis with a statistical p-value=0.01. Trapping Time was about 2% in the case of controls and remained substantially unchanged for osteoporosis. Entropy was about 0.300-1.100 bits in the case of controls but assumed values about 1.100-1.400 bits in the case of osteoporosis.

Table 1. Recurrences Quantification Analysis of 99mTc-HDP Nuclear Image of Bone for Oral Diagnosis of Osteoporosis

Controls

Subject	% Recurrences	% Determinism	% Laminarity	Trapping Time	Ratio	Entropy	Max Line	Age (years)	Osteoporosis
RED									
Cap.-IR	2.265	12.468	19.141	2.178	5.507	0.419	7	73	no
Pag.-IR	0.626	12.049	17.502	2.136	19.256	0.442	7	57	no
Sch.-IR	2.218	16.336	23.951	2.238	7.366	0.544	8	57	no
Deb.-IR	1.453	13.791	21.022	2.221	9.494	0.473	9	73	no
Tal.-IR	1.979	12.539	17.493	2.128	6.337	0.393	8	68	no
m.v.	1.708	13.437	19.822	2.180	9.592	0.454	7.800		
st.dev	0.685	1.747	2.728	0.049	5.604	0.058	0.837		
GREEN									
Cap.-IR	2.123	12.329	19.162	2.172	5.807	0.414	7		
Pag.-IR	0.767	11.286	15.790	2.144	14.722	0.423	6		
Sch.-IR	1.878	15.875	23.026	2.241	8.454	0.253	8		
Deb.-IR	1.514	13.723	20.750	2.231	9.055	0.470	7		
Tal.-IR	1.293	12.146	17.579	2.125	9.392	0.380	7		
m.v.	1.515	13.072	19.261	2.183	9.486	0.388	7.000		
st.dev	0.527	1.794	2.796	0.052	3.248	0.082	0.707		
BLEU									
Cap.-IR	3.472	12.816	19.562	2.174	3.692	0.429	6		
Pag.-IR	2.753	11.753	17.241	2.156	4.270	0.391	6		
Sch.-IR	3.204	15.617	22.321	2.225	4.874	0.534	8		
Deb.-IR	3.747	13.594	19.991	2.184	3.628	0.464	7		
Tal.-IR	2.896	12.416	17.378	2.137	4.287	3.840	7		
m.v.	3.214	13.239	19.299	2.175	4.150	1.132	6.800		
st.dev	0.408	1.487	2.098	0.033	0.510	1.515	0.837		

Table 1. (Continued)

Subject	% Recurrences	% Determinism	% Laminarity	Trapping Time	Ratio	Entropy	Max Line	Age (years)	Osteoporosis
RED									
Cap.-IL	4.138	16.227	26.612	2.233	3.921	0.568	9		
Pag.-IL	2.655	14.619	21.720	2.180	5.506	0.438	10		
Sch.-IL	2.658	13.639	20.343	2.220	5.131	0.496	7		
Deb.-IL	2.352	16.450	23.812	2.253	6.995	0.575	9		
Tal.-IL	5.009	14.057	20.627	2.180	2.657	0.475	8		
m.v.	3.362	14.998	22.623	2.213	4.842	0.510	8.600		
st.dev	1.154	1.274	2.614	0.033	1.642	0.060	1.140		
GREEN									
Cap.-IL	2.097	15.215	24.336	2.250	7.257	0.541	8		
Pag.-IL	1.893	14.712	22.101	2.180	7.815	0.452	6		
Sch.-IL	2.598	13.520	20.548	2.228	5.204	0.492	7		
Deb.-IL	1.838	16.966	24.082	2.262	9.229	0.591	10		
Tal.-IL	3.379	13.186	19.305	2.165	3.902	0.436	7		
m.v.	2.361	14.720	22.074	2.217	6.681	0.502	7.600		
st.dev	0.643	1.507	2.188	0.043	2.122	0.064	1.517		
BLEU									
Cap.-IL	2.266	14.664	22.952	2.190	6.471	0.508	8		
Pag.-IL	2.270	13.487	19.885	2.124	5.941	0.419	7		
Sch.-IL	3.985	13.880	21.067	2.200	3.483	0.496	7		
Deb.-IL	2.701	16.573	23.389	2.214	6.137	0.566	9		
Tal.-IL	2.422	12.487	18.234	2.151	5.134	0.410	7		
m.v.	2.729	14.218	21.105	2.176	5.433	0.480	7.600		
st.dev	0.724	1.532	2.142	0.037	1.196	0.065	0.894		

Subject	% Recurrences	% Determinism	% Laminarity	Trapping Time	Ratio	Entropy	Max Line	Age (years)	Osteoporosis
RED									
Cap.-SR	0.875	8.872	12.776	2.083	10.143	0.301	5		
Pag.-SR	0.836	13.023	20.251	2.241	15.571	0.483	7		
Sch.-SR	0.803	14.508	21.700	2.227	18.075	0.541	12		
Deb.-SR	1.713	13.534	19.756	2.174	7.901	0.445	7		
Tal.-SR	0.891	9.458	14.114	2.078	10.611	0.303	5		
m.v.	1.024	11.879	17.719	2.161	12.460	0.415	7.200		
st.dev	0.387	2.543	3.995	0.077	4.206	0.108	2.864		
GREEN									
Cap.-SR	0.885	8.536	12.421	2.088	9.642	0.294	6		
Pag.-SR	0.908	11.334	18.630	2.192	12.486	0.405	6		
Sch.-SR	0.930	13.654	21.565	2.231	14.688	0.507	12		
Deb.-SR	1.639	13.360	19.626	2.168	8.151	0.440	6		
Tal.-SR	1.015	9.425	13.371	2.077	9.285	0.306	5		
m.v.	1.075	11.262	17.123	2.151	10.850	0.390	7.000		
st.dev	0.319	2.288	4.014	0.067	2.674	0.090	2.828		
BLEU									
Cap.-SR	3.261	9.127	12.276	2.083	2.799	0.293	6		
Pag.-SR	2.759	11.935	19.552	2.191	4.326	0.401	6		
Sch.-SR	2.010	13.223	20.784	2.217	6.578	0.486	12		
Deb.-SR	3.653	13.372	19.081	2.167	3.660	0.441	7		
Tal.-SR	2.358	9.551	13.486	2.079	4.050	0.301	6		
m.v.	2.808	11.442	17.036	2.147	4.283	0.384	7.400		
st.dev	0.664	2.005	3.867	0.063	1.407	0.085	2.608		

Table 1. (Continued)

Subject	% Recurrences	% Determinism	% Laminarity	Trapping Time	Ratio	Entropy	Max Line	Age (years)	Osteoporosis
RED									
Cap.-SL	1.827	10.125	15.479	2.112	5.543	0.337	7		
Pag.-SL	1.861	12.667	18.883	2.112	6.808	0.399	7		
Sch.-SL	0.765	11.533	18.195	2.164	15.086	0.444	9		
Deb.-SL	2.728	14.150	21.179	2.193	5.187	0.444	8		
Tal.-SL	1.196	11.959	18.916	2.154	9.982	0.376	6		
m.v.	1.675	12.087	18.530	2.147	8.521	0.400	7.400		
st.dev	0.746	1.480	2.044	0.035	4.128	0.046	1.140		
GREEN									
Cap.-SL	1.808	10.074	15.375	2.109	5.572	0.336	7		
Pag.-SL	1.431	12.089	18.301	2.112	8.450	0.382	7		
Sch.-SL	0.752	10.446	16.941	2.138	13.889	0.406	9		
Deb.-SL	1.506	14.134	22.861	2.210	9.385	0.467	8		
Tal.-SL	1.145	11.950	18.715	2.157	8.448	0.373	6		
m.v.	1.328	11.739	18.439	2.145	9.149	0.393	7.400		
st.dev	0.399	1.608	2.796	0.041	3.012	0.049	1.140		
BLEU									
Cap.-SL	3.437	10.413	16.503	2.098	3.029	0.342	6		
Pag.-SL	2.695	11.949	16.977	2.143	4.433	0.371	6		
Sch.-SL	2.291	11.624	18.157	2.126	5.073	0.389	9		
Deb.-SL	2.337	13.707	20.515	2.163	5.866	0.447	8		
Tal.-SL	3.106	11.737	18.142	2.137	3.892	0.384	7		
m.v.	2.773	11.886	18.059	2.133	4.459	0.387	7.200		
st.dev	0.495	1.181	1.552	0.024	1.087	0.038	1.304		

R.O.I.: Maxillofacial bone; IR and IL state for lower right and lower left, respectively; SR and SL state for upper right and upper left, respectively.

Table 2. Recurrences Quantification Analysis of ^{99m}Tc-HDP Nuclear Image of Bone for Oral Diagnosis of Osteoporosis

		Patients with osteoporosis							
Subject	% Recurrences	% Determinism	% Laminarity	Trapping Time	Ratio	Entropy	Max Line	Age (years)	Osteoporosis
RED									
Mas.-IR	3.006	50.697	85.273	2.406	16.855	0.829	191	43	+ + -
Dem.-IR	8.900	60.390	77.913	3.658	6.785	1.759	67	59	+ + +
Pan.-IR	2.962	43.716	61.696	3.107	14.757	1.455	58	60	+ + +
Ceg.-IR	12.999	75.057	88.687	4.825	5.774	2.301	231	63	+ + +
Gre.-IR	4.391	18.529	27.708	2.262	4.220	0.602	10	48	+ - -
m.v.	6.452	49.678	68.255	3.252	9.678	1.389	111.400		
st.dev	4.392	21.010	24.937	1.044	5.716	0.690	94.532		
GREEN									
Mas.-IR	1.533	49.838	85.512	2.365	32.519	0.785	191		
Dem.-IR	4.842	64.152	78.774	4.036	13.249	2.004	69		
Pan.-IR	1.657	38.303	54.777	3.027	23.120	1.425	58		
Ceg.-IR	7.840	73.343	88.083	4.838	9.355	2.262	203		
Gre.-IR	2.121	17.675	25.606	2.260	8.334	0.592	8		
m.v.	3.599	48.662	66.550	3.305	17.315	1.414	105.800		
st.dev	2.727	21.897	26.399	1.111	10.313	0.731	86.474		
BLEU									
Mas.-IR	2.647	77.613	84.639	2.308	17.984	0.733	191		
Dem.-IR	1.553	38.927	55.044	3.444	25.064	1.539	69		
Pan.-IR	2.622	29.446	44.451	2.807	11.232	1.137	58		
Ceg.-IR	3.163	65.837	85.463	4.070	20.816	1.931	201		
Gre.-IR	2.143	16.202	24.267	2.220	7.560	0.537	8		
m.v.	2.426	45.605	58.773	2.970	16.531	1.175	105.400		
st.dev	0.607	25.515	26.415	0.785	7.104	0.572	85.914		

Table 2. (Continued)

Subject	% Recurrences	% Determinism	% Laminarity	Trapping Time	Ratio	Entropy	Max Line	Age (years)	Osteoporosis
RED									
Mas.-IL	4.316	12.826	19.109	2.141	2.972	0.443	8		
Dem.-IL	9.326	50.852	69.673	3.407	5.453	1.501	49		
Pan.-IL	9.174	53.678	74.388	4.041	5.851	1.681	73		
Ceg.-IL	11.678	79.707	90.053	5.888	6.825	2.588	150		
Gre.-IL	4.185	19.651	28.442	2.375	4.690	0.676	11		
m.v.	7.736	43.343	56.333	3.570	5.158	1.378	58.200		
st.dev	3.333	27.286	30.840	1.508	1.444	0.857	58.049		
GREEN									
Mas.-IL	2.134	12.025	18.114	2.152	5.635	0.412	8		
Dem.-IL	8.510	58.693	75.881	3.830	6.897	1.746	49		
Pan.-IL	3.576	57.023	75.500	3.873	15.945	1.756	72		
Ceg.-IL	15.023	83.892	93.996	6.938	5.584	2.739	150		
Gre.-IL	2.121	19.446	29.045	2.409	9.169	0.606	10		
m.v.	6.273	46.216	58.507	3.840	8.646	1.452	57.800		
st.dev	5.550	29.907	32.976	1.904	4.332	0.953	58.191		
BLEU									
Mas.-IL	2.753	11.820	17.664	2.128	4.929	0.393	8		
Dem.-IL	2.439	42.143	61.826	3.584	17.280	1.521	49		
Pan.-IL	1.878	38.713	61.115	3.364	20.617	1.368	72		
Ceg.-IL	7.561	84.346	94.581	7.623	11.155	2.960	150		
Gre.-IL	2.220	18.480	27.621	2.335	8.326	0.626	10		
m.v.	3.370	39.100	52.561	3.807	12.461	1.374	57.800		
st.dev	2.364	28.399	30.678	2.225	6.428	1.007	58.191		

Subject	% Recurrences	% Determinism	% Laminarity	Trapping Time	Ratio	Entropy	Max Line	Age (years)	Osteoporosis
RED									
Mas.-SR	2.621	28.145	41.326	2.600	10.739	0.940	193		
Dem-SR	4.512	75.558	98.825	3.075	16.745	1.253	411		
Pan.-SR	1.546	29.058	45.696	2.796	18.791	1.115	35		
Ceg.-SR	12.994	72.589	87.611	4.708	5.586	2.203	123		
Gre.-SR	0.702	13.159	19.519	2.168	18.734	0.491	9		
m.v.	4.475	43.702	58.595	3.069	14.119	1.200	154.200		
st.dev	4.971	28.454	33.361	0.974	5.793	0.630	161.025		
GREEN									
Mas.-SR	1.248	24.448	34.078	2.430	19.597	0.868	190		
Dem-SR	1.873	75.905	98.064	3.133	40.517	1.415	411		
Pan.-SR	1.083	21.867	36.139	2.640	20.199	0.974	35		
Ceg.-SR	5.423	75.375	87.680	4.420	13.898	2.235	120		
Gre.-SR	0.833	12.739	19.702	2.170	15.298	0.480	9		
m.v.	2.092	42.067	55.133	2.959	21.902	1.194	153.000		
st.dev	1.901	30.956	35.220	0.890	10.752	0.670	160.998		
BLEU									
Mas.-SR	2.488	21.534	32.394	2.379	8.654	0.779	190		
Dem-SR	1.883	68.029	98.033	2.593	36.134	0.942	411		
Pan.-SR	2.506	18.040	36.505	2.550	7.199	0.694	35		
Ceg.-SR	2.563	63.324	80.679	4.083	24.711	1.985	117		
Gre.-SR	2.164	12.380	18.903	2.153	5.720	0.430	9		
m.v.	2.321	36.661	53.303	2.752	16.484	0.966	152.400		
st.dev	0.290	26.739	34.106	0.764	13.390	0.599	161.158		

Table 2. (Continued)

Subject	% Recurrences	% Determinism	% Laminarity	Trapping Time	Ratio	Entropy	Max Line	Age (years)	Osteoporosis
RED									
Mas.-SL	5.143	12.296	17.722	2.122	2.391	0.393	186		
Dem.-SL	2.815	40.602	58.376	3.172	14.425	1.239	94		
Pan.-SL	6.622	50.395	74.708	4.469	7.611	1.830	86		
Ceg.-SL	11.797	79.128	90.721	5.738	6.707	2.504	174		
Gre.-SL	0.780	14.568	21.932	2.249	18.681	0.574	10		
m.v.	5.431	39.398	52.692	3.550	9.963	1.308	110.000		
st.dev	4.198	27.622	32.141	1.542	6.507	0.878	71.944		
GREEN									
Mas.-SL	3.482	11.261	16.699	2.129	3.254	0.356	186		
Dem.-SL	1.438	39.353	55.591	2.982	27.365	1.252	47		
Pan.-SL	3.328	55.226	75.747	4.384	16.595	1.932	86		
Ceg.-SL	15.933	80.217	91.138	5.678	5.035	2.545	174		
Gre.-SL	0.926	13.480	20.311	2.211	14.555	0.538	8		
m.v.	5.021	39.907	51.897	3.477	13.361	1.325	100.200		
st.dev	6.203	29.064	33.011	1.527	9.740	0.925	78.008		
BLEU									
Mas.-SL	2.008	10.629	16.426	2.104	5.294	0.347	187		
Dem.-SL	2.158	30.641	46.373	2.648	14.199	0.974	17		
Pan.-SL	1.804	41.938	63.974	3.879	23.241	1.764	86		
Ceg.-SL	7.992	82.896	91.973	6.104	10.372	2.711	175		
Gre.-SL	2.087	13.167	20.538	2.185	6.310	0.481	10		
m.v.	3.210	35.854	47.857	3.384	11.883	1.255	95.000		
st.dev	2.677	29.275	31.394	1.678	7.265	0.985	84.045		

Statistical analysis was performed using t-test for all seven RQA variables and colours (R, G, B) in controls against subjects with osteoporosis. Statistical significance was obtained for the two groups of subjects with p values ranging from 0.0441 to 0.0026.

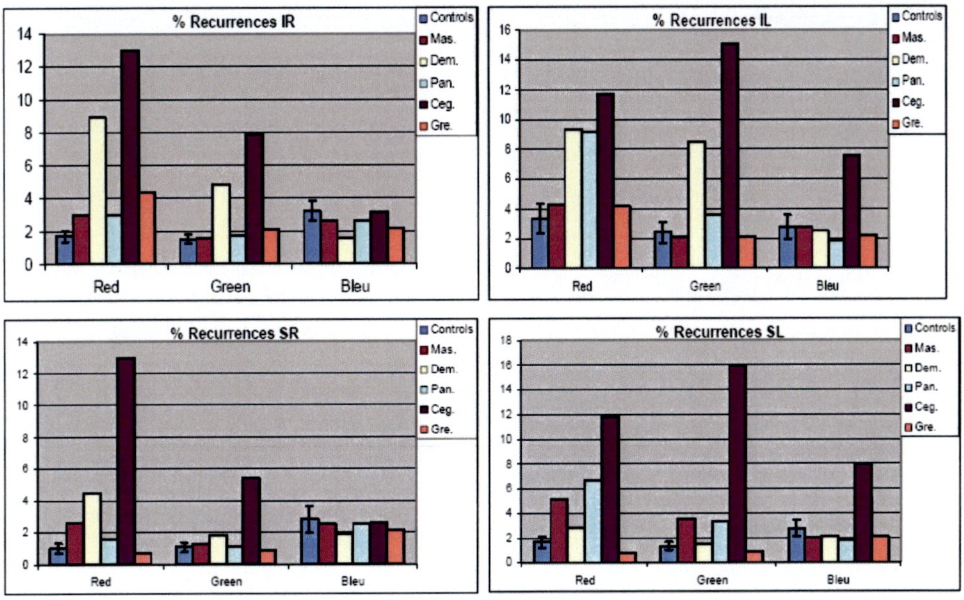

Figure 13.

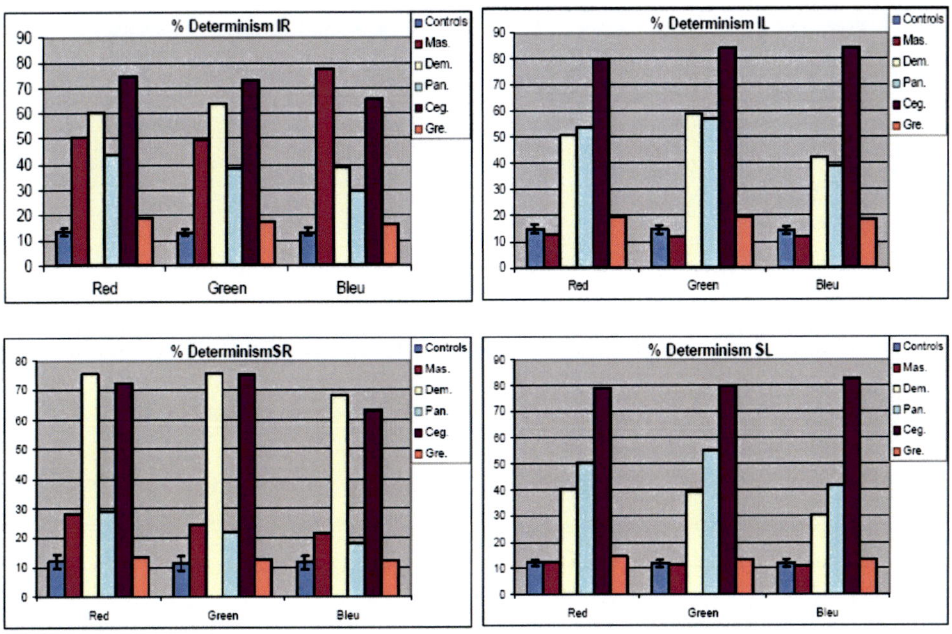

Figure 14.

Statistical p-value resulted p=0.02. Very significant resulted also the differences for controls respect to osteoporosis in the case of Max Line. We repeat here that its inverse gives a measure of the divergence or, as to say, of the chaoticity of the process. In controls, Max Line assumed values ranging from 6.800 to 7.400 while in the case of osteoporosis Max Line assumed values from about 58 to 150. The significance of the statistical results resulted given with p values about 0.002.. Details of the obtained results are given in any particular in Tables

1 and 2 and in Figures 13, 14, 15, and 16. In conclusion, the inspection of the obtained results evidence the same tendency: all the variables values of the RQA differentiate in a net way controls from osteoporosis, and indicate a net difference in the mechanism of bone uptake of ^{99m}Tc-labeled diphosphonates in the case of controls respect to subjects with osteoporosis. In addition, looking at Figures 13, 14, 15, and 16 and Table 2 a net correlation may be also found between the values assumed from variables in RQA and the level of severity of the pathology.

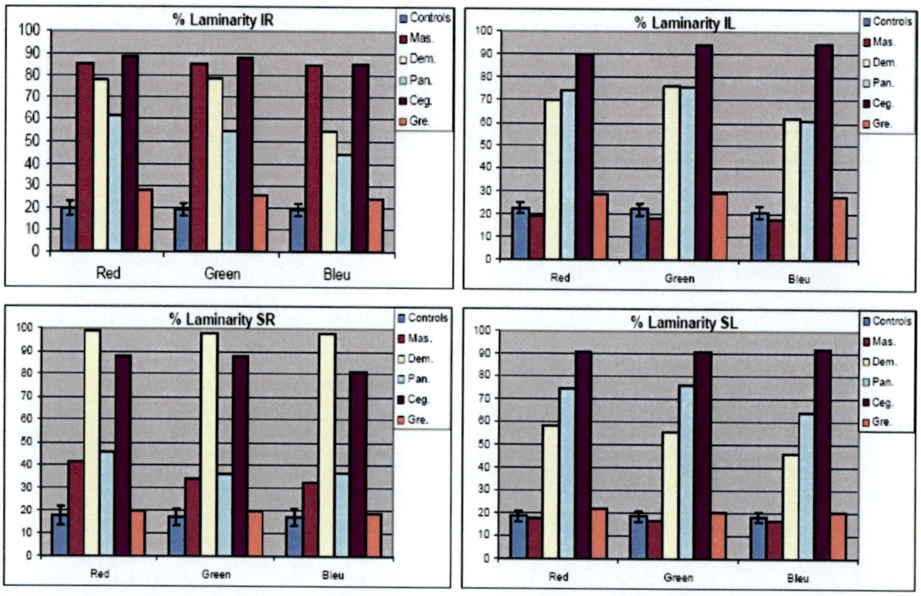

Figure 15.

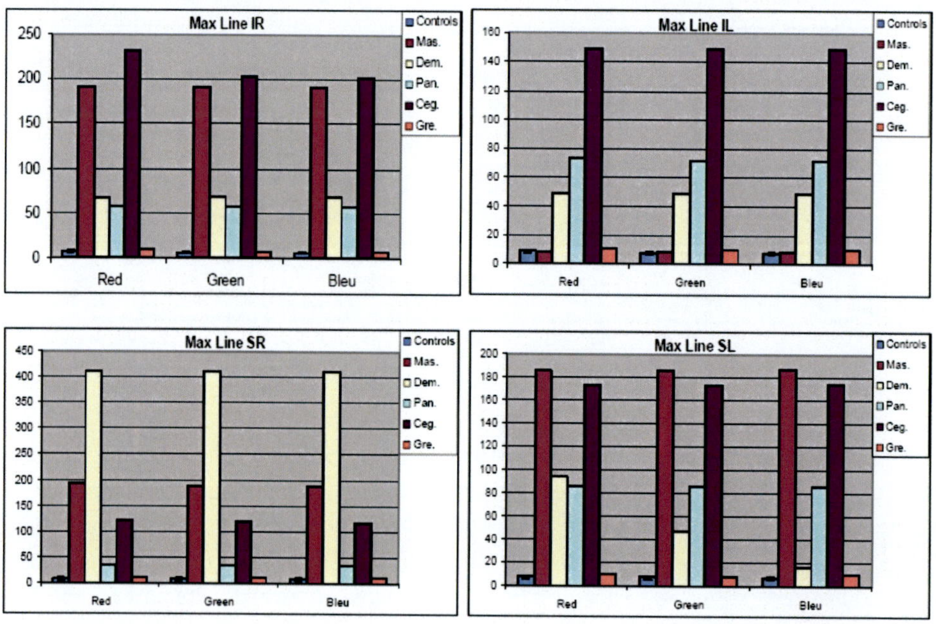

Figure 16.

Table 3. Chaos Analysis: estimation of Fractal Dimension by Hurst exponent

Controls	H (Hurst Exponent)				D (Fractal Dimension D=2-H)			
Subject	IR	IL	SR	SL	IR	IL	SR	SL
RED								
Cap.	0.2916	0.2195	0.1298	0.3702	1.7084	1.7805	1.8702	1.6298
Pag.	0.1835	0.2418	0.3405	0.2668	1.8165	1.7582	1.6595	1.7332
Sch.	0.3196	0.2726	0.4577	0.3827	1.6804	1.7274	1.5423	1.6173
Deb.	0.5610	0.3499	0.3629	0.2111	1.4390	1.6501	1.6371	1.7889
Tal.	0.2829	0.3195	0.5687	0.5839	1.7171	1.6805	1.4313	1.4161
m.v.	0.3277	0.2807	0.3719	0.3629	1.6723	1.7193	1.6281	1.6371
st.dev	0.1402	0.0539	0.1626	0.1428	0.1402	0.0539	0.1626	0.1428
GREEN								
Cap.	0.3047	0.3064	0.4094	0.3326	1.6953	1.6936	1.5906	1.6674
Pag.	0.4152	0.3025	0.3655	0.2669	1.5848	1.6975	1.6345	1.7331
Sch.	0.3028	0.2462	0.4209	0.4191	1.6972	1.7538	1.5791	1.5809
Deb.	0.4704	0.4114	0.3126	0.2519	1.5296	1.5886	1.6874	1.7481
Tal.	0.4414	0.4587	0.4513	0.4632	1.5586	1.5413	1.5487	1.5368
m.v.	0.3869	0.3450	0.3919	0.3467	1.6131	1.6550	1.6081	1.6533
st.dev	0.0784	0.0872	0.0540	0.0927	0.0784	0.0872	0.0540	0.0927
BLEU								
Cap.	0.2283	0.2986	0.1298	0.1682	1.7717	1.7014	1.8702	1.8318
Pag.	0.2385	0.2900	0.2211	0.1955	1.7615	1.7100	1.7789	1.8045
Sch.	0.2398	0.1833	0.3214	0.2418	1.7602	1.8167	1.6786	1.7582
Deb.	0.1810	0.3182	0.1883	0.2533	1.8190	1.6818	1.8117	1.7467
Tal.	0.3638	0.3839	0.2980	0.3135	1.6362	1.6161	1.7020	1.6865
m.v.	0.2503	0.2948	0.2317	0.2345	1.7497	1.7052	1.7683	1.7655
st.dev	0.0679	0.0724	0.0788	0.0560	0.0679	0.0724	0.0788	0.0560

Table 4. Chaos Analysis: estimation of Fractal Dimension by Hurst exponent

Patients with osteoporosis								
	H (Hurst Exponent)				D (Fractal Dimension D=2-H)			
Subject	IR	IL	SR	SL	IR	IL	SR	SL
RED								
Mas.	0.3927	0.1286	0.5122	0.0721	1.6073	1.8714	1.4878	1.9279
Dem.	0.3266	0.3810	0.4671	0.5345	1.6734	1.6190	1.5329	1.4655
Pan.	0.3619	0.3058	0.5367	0.2901	1.6381	1.6942	1.4633	1.7099
Ceg.	0.1754	0.2671	0.2332	0.2847	1.8246	1.7329	1.7668	1.7153
Gre.	0.2649	0.0996	0.4784	0.5206	1.7351	1.9004	1.5216	1.4794
m.v.	0.3043	0.2364	0.4455	0.3404	1.6957	1.7636	1.5545	1.6596
st.dev	0.0863	0.1194	0.1218	0.1922	0.0863	0.1194	0.1218	0.1922
GREEN								
Mas.	0.4166	0.2394	0.4855	0.1595	1.5834	1.7606	1.5145	1.8405
Dem.	0.3995	0.5117	0.4632	0.5114	1.6005	1.4883	1.5368	1.4886
Pan.	0.4326	0.4963	0.4414	0.3672	1.5674	1.5037	1.5586	1.6328
Ceg.	0.2678	0.3819	0.5032	0.2311	1.7322	1.6181	1.4968	1.7689
Gre.	0.2990	0.2044	0.3647	0.4201	1.7010	1.7956	1.6353	1.5799
m.v.	0.3631	0.3667	0.4516	0.3379	1.6369	1.6333	1.5484	1.6621
st.dev	0.0745	0.1419	0.0539	0.1423	0.0745	0.1419	0.0539	0.1423
BLEU								
Mas.	0.3399	0.2154	0.3525	0.2541	1.6601	1.7846	1.6475	1.7459
Dem.	0.4716	0.4360	0.4280	0.4062	1.5284	1.5640	1.5720	1.5938
Pan.	0.3449	0.4382	0.2374	0.3599	1.6551	1.5618	1.7626	1.6401
Ceg.	0.4227	0.4989	0.4533	0.3908	1.5773	1.5011	1.5467	1.6092
Gre.	0.1934	0.2503	0.2320	0.2091	1.8066	1.7497	1.7680	1.7909
m.v.	0.3545	0.3678	0.3406	0.3240	1.6455	1.6322	1.6594	1.6760
st.dev	0.1056	0.1263	0.1036	0.0875	0.1056	0.1263	0.1036	0.0875

At the third point of our investigation, we estimated fractal dimension calculating the Hurst exponent. We will not enter in the details of the calculation here since the methodology is well known and it is well exposed in basic papers in literature [9]. The results are given in Tables 3 and 4.

Statistical controls:

SR green (controls vs patients with osteoporosis): significance p=0.1184
SR bleu (controls vs patients with osteoporosis): significance p=0.0982
IR bleu (controls vs patients with osteoporosis): significance p=0.1005
SL bleu (controls vs patients with osteoporosis): significance p=0.0900

Two way Anova

IR (controls vs patients with osteoporosis): significance p=0.0842
IL (controls vs patients with osteoporosis): significance p=0.0005
SR (controls vs patients with osteoporosis): significance p=0.0519
SL (controls vs patients with osteoporosis): significance p=0.0254

The first conclusion to which one arrives by inspection of such results is that the mechanism of adsorption of phosphate compounds with ^{99m}Tc to amorphous bone is a fractal process. Therefore, it is a rather complex mechanism that may be investigated correctly only using non linear methods of analysis as the RQA and the variability analysis that we previously considered.

The second conclusion is that the estimation of fractal dimension in subjects with osteoporosis reveals also in this case that we are in presence of a profound alteration and deterioration of the same bone uptake mechanism from this radiopharmaceutical .Statistical analysis and, in particular, the use of ANOVA indicates significant differences in fractal dimension with p-values ranging from 0.1184 to 0.0005, depending on the ROI investigated. Therefore, osteoporosis induces a strong dismantling in the bone uptake mechanism of ^{99m}Tc-labeled diphosphonates and the use of fractal dimension represents an adequate mandibular or, in general, maxillary bone index in order to account with rigour and accuracy on the level of severity of the pathology under examination.

Conclusion

We would summarize briefly some of the results obtained in this paper.

The earliest suggestion on an association between osteoporosis and oral bone loss dates back to 1960 [10]. Dental radiographs represent the most frequently used tools in this professional sphere of competence. However, in this paper we have suggested that the use of maxillary bone ^{99m}Tc-labeled diphosphonates imaging may represent an extremely valid technique in order to analyze osteoporosis and alterations of such bone districts. A very important feature is that by using ^{99m}Tc-labeled diphosphonates imaging we are able to develop a micro structural and micro architectural analysis that informs us on the dynamic processes that are happening at these levels of observation. In addition, the use of ROI technique enables us in general to select in any individual case the collocation of the ROI and its dimension. The possibility of such procedure allow the dentist to evaluate with high accuracy and detail, the oral conditions of alteration that relate his intervention. Finally, one must observe that, in order to quantify the level of severity of the oral pathology, some indices should be employed. To this purpose one must take care, acknowledging that osteoporosis, characterized by a loss of bone mineral density, is a dynamic process highly non linear. Mandibular or, generally, maxillary bones indices must account with rigour of such basic physiologic condition. Therefore, such indices must be the result of non linear methodologies. The use of variability, of RQA and of Fractal Dimension estimation result absolutely indispensable in this case.

REFERENCES

[1] R.I. Ferreira, S.M. de Almeida, N. Boscolo, A.O. Santos, E.E. Camargo, Bone Scintigraphy as Adjunct for the Diagnosis of oral diseases, *Journal of Dental Education*, 66,12,1381-1387, 2002.

[2] G. Subramanian, Radiopharmaceuticals for bone scanning, *Skeletal Nuclear Medicine*, St.Louis, Mosby, 9, 20,1996.

[3] E. Conte, M. Mele, A. Fratello, D. Pasculli, M. Pieralice, A. D'Addabbo, Computer Analysis of Yc-99mDPD and Tc-99mMDP Kinetics in Human, Journal Nuclear Medicine, 24, 334-338, 1983 and references therein; M. Mastrolonardo, E. Conte, J.P. Zbilut, A Fractal Analysis of Skin Pigmented Lesions using the Novel Tool of the Variogram Technique, *Chaos, Solitons and Fractals*, 28, 5, 1119-1135, 2006.

[4] C.L. Webber, J.P. Zbilut, Dynamical Assessment of Physiological Systems and States using Recurrence Plot Strategies, J. Appl. Physiol. 76, 965-973, 1994; J.P. Zbilut, N. Thomasson, CL. Webber, Recurrence Quantification Analysis as a tool for non linear exploration of non stationary cardiac signals, *Medical. Eng. Phys.*, 24, 53-60, 2002; N. Marwan, Encounters with Neighbours – Dissertation Universitat Potsdam Institut für Physik, Mai 2003.

[5] For some applications, see E. Conte, A. Vena, A. Federici, R. Giuliani, J.P. Zbilut, A brief note on possible detection of physiological singularities in respiratory dynamics by recurrence quantification analysis of lung sounds, *Chaos, Solitons and Fractals*, 21, 869-877, 2004.

[6] E. Conte, A. Federici, M. Minervini, A. Papagni, and J.P. Zbilut, Measures of Coupling strength and synchronization in non linear interaction of heart rate and systolic blood pressure in the cardiovascular control system, 2, 1, 1-22, 2006. For a detailed exposition of principles and applications see: J.P. Zbilut, A. Giuliani, Simplicity: *The Latent Order of Complexity,* Nova Publishers, 2007.

[7] J.P. Eckmann, S.O. Kamphorst, D. Ruelle, Recurrence Plot of Dynamical Systems, *Europhys. Letters*, 4, 973-977, 1987.

[8] F. Yasar, F. Akgunlu, The Differences in Panoramics Mandibular Indices and Fractal Dimension between Patients with and without spinal osteoporosis, *Dentomaxillofacial Radiology*, 35, 1-9, 2006.

[9] L. Pothuaud, E. Lespessailles, R. Harba, R. Jennane, V. Royant, E. Eynard, C.L. Benhamou, Fractal Analysis of Trabecular Bone Texture on Radiographs: Discriminant

Value in Postmenopausal Osteoporosis, *Osteoporosis International*, 8, 618-625,1998 and references therein.

[10] H.E. Hurst, P.Black, Y.M. Sinaika, Long-Term Storage in Reservoirs: An Experimental Study, Constable, London, 1965; J.J. Groen, F. Duyvensz, J. A. Halsted, Diffuse Alveolar Atrophy of the Jaw (non infiammatory form of paradental disease) and presenile osteoporosis, *Geront. Clin.* 2, 68-86, 1960.

In: Progress in Chaos Complexity Research
Editors: Franco F. Orsucci and Nicoletta Sala
ISBN: 978-1-60456-375-7

Chapter 9

FORECASTING OF HYPERCHAOTIC RÖSSLER SYSTEM STATE VARIABLES USING ONE OBSERVABLE

Massimo Camplani and Barbara Cannas
Electrical and Electronic Engineering Dept.
University of Cagliari, Italy

ABSTRACT

In the last years, growing attention has been paid to the reconstruction of chaotic attractors from one or more observables. In this paper a Multi Layer Perceptron with a tapped line as input, is used to forecast the hypercaotic Rössler system state variables starting from measurements of one observable. Results show satisfactory prediction performance if a sufficient number of taps is used. Moreover, a sensitivity analysis has been performed to evaluate the predictiveness of the different delayed input in the neural network model.

Keywords: *Multi Layer Perceptron, Forecasting, Embedding dimension, Hyperchaos.*

1. INTRODUCTION

Nonlinear time series forecasting is one of the main tasks of nonlinear time series analysis and it is studied in several application domains such as economy, inventory and production control, weather, signal processing, hydrology, fluid flow, etc. In the last twenty years, growing attention has been paid to the reconstruction of chaotic attractors from one or more observables. This is a challenging topic since it allows one to test the accuracy of the predictive models in case of time series characterized by apparently random behaviour. In fact, chaotic time series appear to be nonperiodic and randomly distributed even if are the output of a completely deterministic process.

Nowadays several neural networks models have been employed to predict chaotic time series. Takens's embedding theorem [1] states that a scalar sequence of measurements $x(t_1)$, $x(t_2)$,.... from a generic dynamic system includes all the information required to completely represent the state space **x**. In particular a scalar d (the embedding dimension), a scalar τ (an arbitrary delay) and a function f exist such that $\mathbf{x}(t) = \mathbf{f}[x(t), x(t-\tau), x(t-2\tau), ..., x(t-d\tau)]$.

The theorem gives a useful lower bound on the minimal number of measurements to obtain the reconstruction of the dynamic system: the embedding dimension d must satisfy $d \geq 2D+1$, where D is the dimension of the attractor. Moreover, there is not an exact embedding dimension which is the best value to reconstruct the phase space and the chaotic attractor. A chaotic time series data can be embedded in phase space of different dimension while preserving the chaotic attractor behaviour.

In this paper, the four state variables of the *hyperchaotic* Rössler system are forecasted with a dynamic neural network fed with samples of only one state variable. At the authors knowledge, no contribution has been proposed in literature regarding the hyperchaotic Rössler attractor. Conversely, several papers present forecasting models of *chaotic* Rössler system state variables. The majority of them concerns the forecasting of some variables from time series of the same variables [2-7]. In [7], the cross-prediction of the x_2 and x_3 variables from samples of x_1 at the same time instant, using externally driven dynamical networks, is presented.

The paper is organized as follows: in Section II the Rössler hyperchaotic system is introduced, Section III reports the results of the neural model and the sensitivity analysis. In Section IV conclusions are drawn.

2. Case Study: The Rössler Hyperchaotic System

An hyperchaotic attractor is typically defined as an attractor with at least two positive Lyapunov exponents, combined with a null one and a negative one. The first hyperchaotic system was presented by Rössler in 1979 [8]. The equations of hyperchaotic Rössler system are:

$$\begin{cases} \dot{x}_1 = -x_2 - x_3 \\ \dot{x}_2 = x_1 + ax_2 + x_4 \\ \dot{x}_3 = x_3 x_3 + b \\ \dot{x}_4 = -cx_3 + dx_4 \end{cases} \qquad (1)$$

The system shows a hyperchaotic behavior for: $a = \frac{1}{4}$, $b = 3$, $c = \frac{1}{2}$, $d = \frac{1}{20}$ and initial conditions: $x_1 = -15$, $x_2 = 11$, $x_3 = 0.2$, $x_4 = 23$. Figures 1(a) and 1(b) show two different 3D views of the attractor .

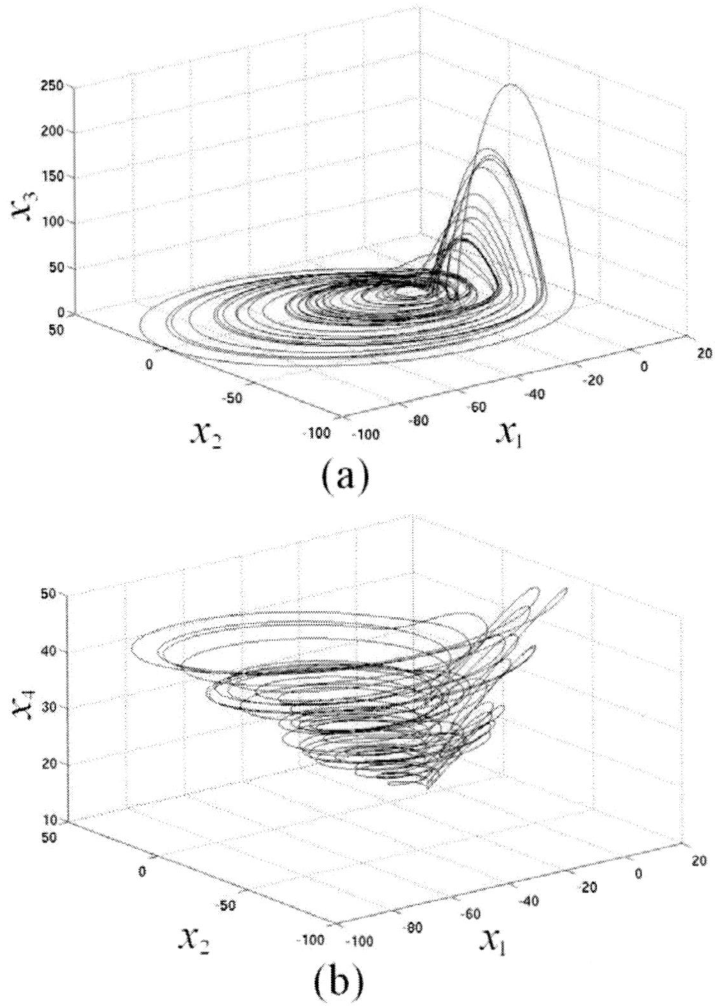

Figure 1. 3D view of the hyperchaotic attractor on the phase space (a) x_1-x_2-x_3 (b) x_1-x_2-x_4.

The reconstruction of the system state from one observable is strictly related to the observability property. The hyperchaotic Rössler system state can be globally reconstructed with respect to all of the state variables. In particular if $x_4(t)$ is known, it is possible to univocally determine the system state, as

$$\begin{cases} x_4 = y \\ x_3 = \dfrac{-\dot{y} + dy}{c} \\ x_1 = \dfrac{\ddot{y} - d\dot{y} - bc}{\dot{y} - dy} \\ x_2 = \dfrac{(\dddot{y} - d\ddot{y})(\dot{y} - dy) - (\ddot{y} - d\dot{y})^2 - bc(\dot{y} - dy)}{(\dot{y} - dy)^2} \end{cases} \tag{2}$$

and consequently it is possible forecasting the system state in t+1.

3. Experimental Results

3.1. The Neural Model

The forecasting of the state variables in t+1 has been performed using a Multi Layer Perceptron (MLP) neural network. An MLP [9] is a machine learning technique which uses a weighted sum of non linear functions for function approximation. It can be used to perform forecasting for nonlinear systems, due to its property of universal approximation [10].

The Rössler hyperchaotic time series are obtained with MATLAB (version 6, release 12) using Runge-Kutta ODE solver with a step size equal to 0.01. The first 12000 samples are considered to build training set (9000 points) and validation set (3000 points). Then, another set of 5000 points is sampled that constitutes an independent test set and it is used to evaluate network performance.

The MLP has as input a tapped delay line of D values of the x_4 state variable $\{x_4(t), x_4(t-\tau), \ldots, x_4[t-(D-1)\tau]\}$, one hidden layer with hyperbolic tangent transfer function and four outputs, $\{x_1(t+1), x_2(t+1), x_3(t+1), x_4(t+1)\}$ (see figure 2).

Several networks have been trained to forecast Rössler system time series, with taps varying from 5 to 110 and hidden neurons varying from 10 to 20. For each configuration, 5 networks with random initial weights have been trained. The best network structure consists of 70 taps and 10 hidden neurons. This architecture is characterized by the minimum Mean Square Error (*MSE*) on the validation set, where

$$MSE = \frac{1}{4} TMSE,$$

and *TMSE* is the Total Mean Square Error, given by $TMSE = \sum_{i=1}^{4} MSE(x_i)$.

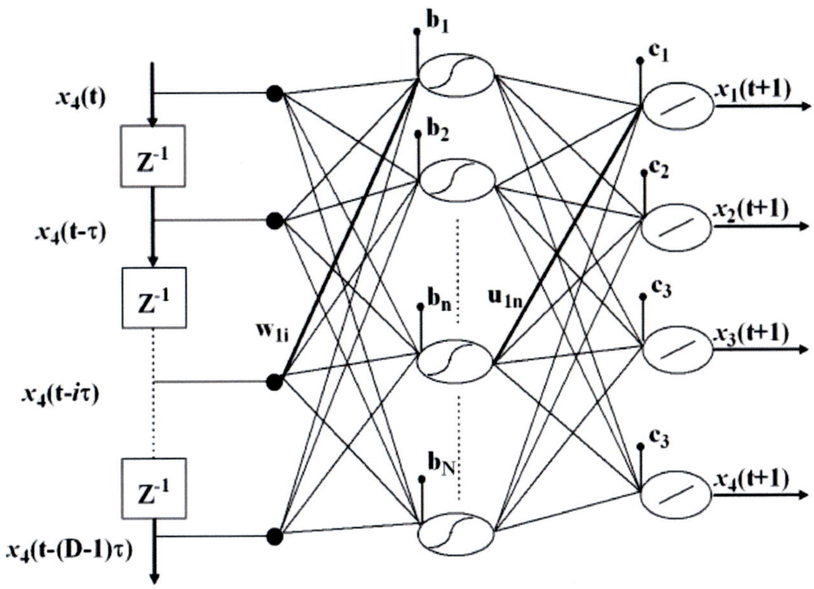

Figure 2. Neural Network scheme.

Figure 3 shows the MSE on the validation data, versus the number of taps for a network with 10 hidden neurons. Red bars represent the minimum MSE (MMSE), obtained in 5 training sessions, with the same network configuration. Blue bars represent the average of the obtained MSEs (AMSEs).

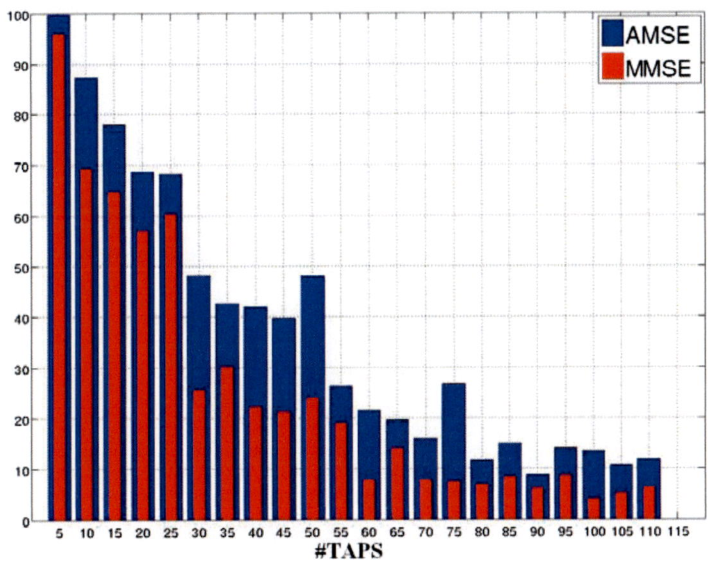

Figure 3. AMSE (Average Mean Square Error, blue bars) and MMSE (Minimum Mean Squared Error, red bars) for different numbers of taps.

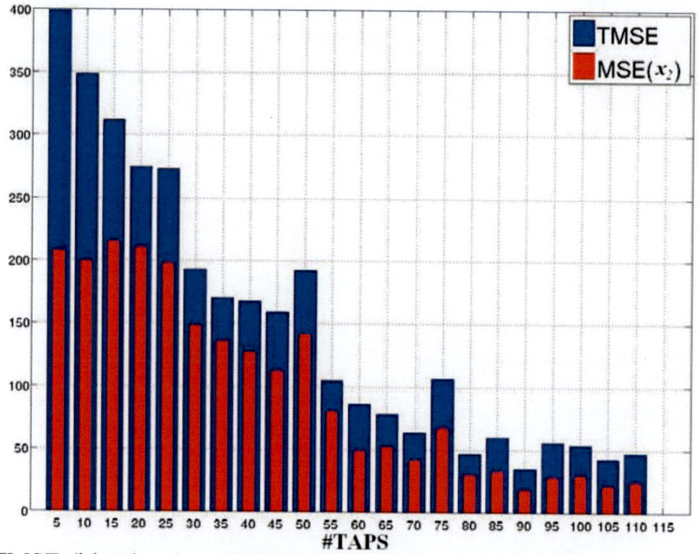

Figure 4. TMSE (blue bars) and MSE(x_2) for networks with different numbers of taps.

The histogram in figure 4 represents the total MSE (blue bars) and the MSE on the x_2 forecasting (red bars). As can be noticed, the error on x_2 is the most influential contribute to the network error. Figure 5 reports the comparison between the network output (dashed line) and the corresponding state variables x_1, x_2, x_3 and x_4. The forecasting shows an increasing error in the variables x_4, x_3, x_1 and x_2. This is due the presence of derivatives of higher order (see eq. (2)) in the relationship between the input x_4(t) and the desired output.

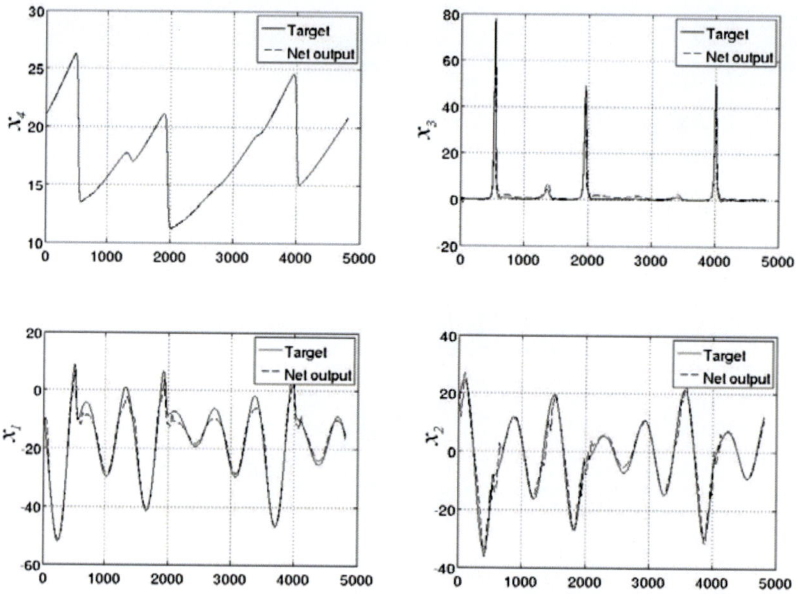

Figure 5. Comparison between network output (dashed line) and target for test data.

Figure 6 shows the error of the best network on the test set.

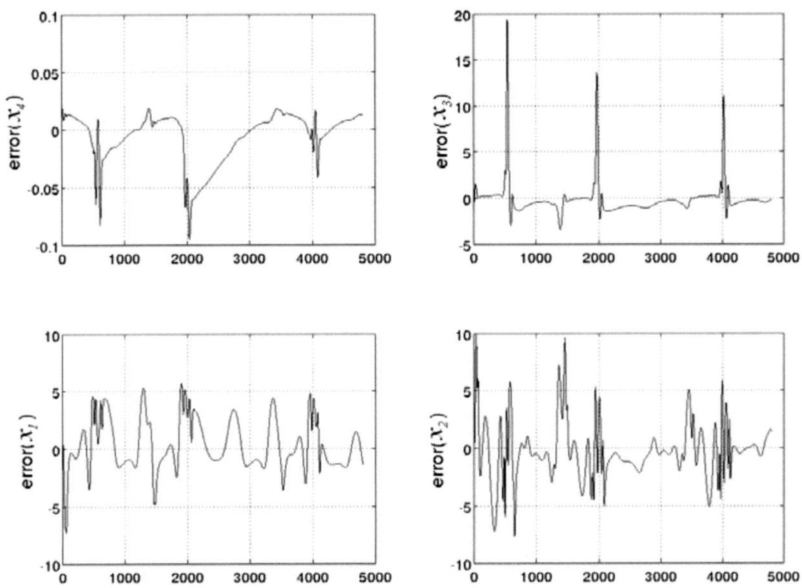

Figure 6. Network error.

3.2. Sensitivity Analysis

Sensitivity analysis is an useful technique to evaluate the neural network model. By calculating the derivatives of the network outputs with respect to the inputs for all patterns, the impact of the independent variables with respect to predicting the dependent variables can be evaluated. Sensitivities identify the magnitude of a relationship between variables but also whether that relationship is positive or negative, conveying a robust instrument of comparison for the evaluation of one independent variable's predictive strength relative to another.

Let us consider the network in figure 2. The jth network output y_j (j = 1,…,4) can be written as follows:

$$y_j = \sum_{n=1}^{N} u_{jn} a_n + c_j = \sum_{n=1}^{N} u_{jn} \tanh\left(\sum_{i=1}^{D} w_{ni} x_i + b_i \right) + c_j \tag{3}$$

where N is the number of hidden neurons, D is the number of inputs (taps), b_n and, w_{ni} are the bias and the weights in the hidden layer, c_j, and u_{jn} are the bias and the weights in the output layer. Appling the chain rule, the sensitivity of the net can be calculated as:

$$\frac{\partial y_j}{\partial x_i} = \sum_{n=1}^{N} u_{jn} \cdot w_{ni} \cdot \frac{\partial}{\partial x_i} \left[\tanh\left(\sum_{i=1}^{D} w_{ni} x_i + b_n \right) \right] \tag{4}$$

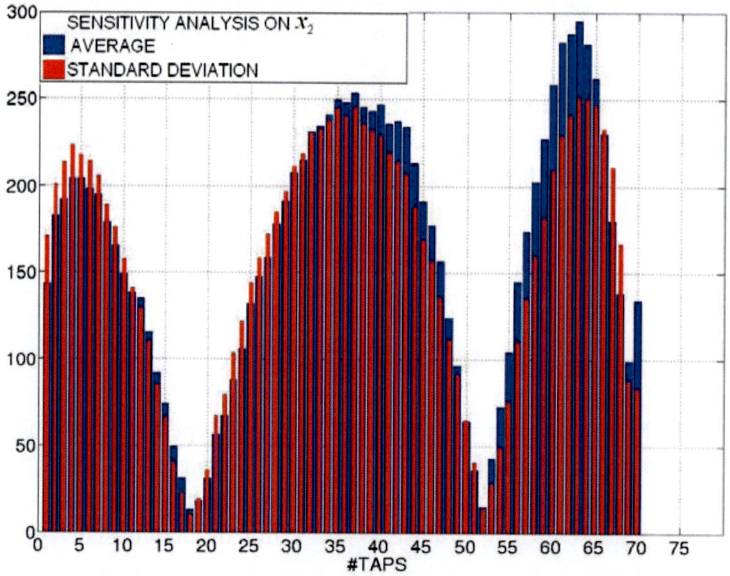

Figure 10. Average sensitivity (blue bar) and standard deviation for the x_2 variable.

A sensistivity analysis has been performed by evaluating the derivatives of the network outputs with respect to the inputs. Results reflect the importance of past inputs in the x_4 and x_3 forecasting.

REFERENCES

[1] Takens F. Detecting strange attractors in turbulence, Dynamical systems and turbulence-Warwick 1980, *LNM 898*, Springer-Verlag.

[2] Han, M.; Fan, M. Application of neural networks on multivariate time series modeling and prediction, *Proc. of American Control Conference,* 2006.

[3] Por, L.T.; Puthusserypady, S. Chaotic time series prediction and additive white Gaussian noise, *Physics Letters* 2007, vol. 365, pp.309-314.

[4] Aoyama, T.; Zhu, H.; Yoshihara, I. Forecasting of the chaos by iterations including multi-layer neural-network, *Proc. of the IEEE-INNS-ENNS International Joint Conference*, 2000, vol. 4, pp. 467-471.

[5] Miyoshi, T.; Ichihashi, H.; Okamoto, S.; Hayakawa, T. Learning chaotic dynamics in recurrent rbf network , in *Proc. of the IEEE International Conference on Neural Networks 1995*, vol. 1, pp. 588-593.

[6] Szpiro, G.G. Forecasting chaotic time series with genetic algorithms, *Phys. Rev.* E. 1997, vol. 55, pp. 2257-2568.

[7] Parlitz, U; Horstein, A. Dynamical prediction of chaotic time series, *Chaos and Complexity Letters* 2005 vol. 1, 135-144.

[8] Rossler, O.E. An equation for hyperchaos, *Physics Letters* 1979, vol, 71, pp.155-157.

[9] Haykin, S.; *Neural Networks, A Comprehensive Foundation.* New York: Macmillan Publishing, 1994.

[10] Cybenko, G. Approximation by superpositions of a sigmoidal function, *Mathematical Control Signal Systems* 1989, vol. 2, pp. 303-314.

In: Progress in Chaos Complexity Research
Editors: Franco F. Orsucci and Nicoletta Sala ISBN: 978-1-60456-375-7

Chapter 10

FRACTAL GEOMETRY IN COMPUTER GRAPHICS AND IN VIRTUAL REALITY

Nicoletta Sala*
[1]Accademia di Architettura,
Università della Svizzera italiana,
Largo Bernasconi,
Mendrisio, Switzerland
[2]Dipartimento di Informatica e Comunicazione
Università dell'Insubria,
Via Mazzini 5, Varese, Italy

ABSTRACT

Fractal geometry is also known as "Mandelbrot's geometry" in honour to its "father" the mathematician Benoit Mandelbrot (b. 1924), that showed how fractals can occur in many different places in mathematics and in other disciplines. Fractal geometry can be used for modelling natural shapes (e.g., fern, trees, seashells, rivers, mountains), and its important applications also appear in computer science, because this "new" geometry permits to compress the images, to reproduce, in the virtual reality environments, the complex patterns and the irregular forms present in nature using simple iterative instructions. The aim of this paper is to present the applications of fractal geometry in computer graphics (to compress images) and in virtual reality (to generate virtual territories and landscapes).

Keywords: *Fractal Geometry, Self-Similarity, Iterated Function Systems, L-Systems, Fractional Brownian Motion, Computer Graphics, Virtual Reality, Virtual World.*

---

* E-mail: nsala@arch.unisi.ch

1. Introduction

French mathematicians Pierre Fatou (1878-1929) and Gaston Julia (1893-1978) worked on fractals at the beginning of the 20th century. The word "Fractal" was coined by the Polish-born Franco-American Benoit Mandelbrot, and it derives from the Latin verb "frangere", "to break", and from the related adjective "fractus", "fragmented and irregular". This term denotes the geometry of nature, which traces inherent order in chaotic shapes and processes, and it was created to differentiate pure geometric figures from other types of figures that defy such simple classification. The Encyclopaedia Britannica introduces the fractal geometry as follow (2007): "In mathematics, the study of complex shapes with the property of self-similarity, known as fractals. Rather like holograms that store the entire image in each part of the image, any part of a fractal can be repeatedly magnified, with each magnification resembling all or part of the original fractal. This phenomenon can be seen in objects like snowflakes and tree bark.... This new system of geometry has had a significant impact on such diverse fields as physical chemistry, physiology, and fluid mechanics; fractals can describe irregularly shaped objects or spatially nonuniform phenomena that cannot be described by Euclidean geometry".

The acceptance of the word "fractal" was dated in 1975, when Mandelbrot presented the list of publications between 1951 and 1975, date when the French version of his book was published, the people were surprised by the variety of the studied fields: economy, linguistics, cosmology, noise on telephone lines, turbulence (Mandelbrot, 1975).

Mandelbrot's geometry replaces Euclidian geometry (which dominated our mathematical thinking for thousands of years) and it is recognized as the true geometry able to describe the Nature. Recently, the electronics evolution and the increase in the computer calculation power permitted connections between the fractal geometry and the other disciplines (for example, biology, economy, medicine, engineering, arts, architecture), and the multiplicity of applications had an important role in the diffusion of fractal geometry (Mandelbrot, 1982; Leland et al., 1993; Nonnenmacker et al., 1994; Eglash, 1999; Barnsley et al., 2002; Sala, 2006; Vyzantiadou et al., 2007). This paper is organized as follows: section 2 introduces the fractal objects, section 3 presents the application of the fractal geometry in computer graphics. Section 4 introduces the fractal geometry as a tool for modelling virtual landscapes. Section 5 is dedicated to the conclusions.

2. Fractal Objects

A fractal object could be defined as a fragmented geometric shape that can be subdivided in parts, each of which is approximately a reduced-size copy of the whole (Mandelbrot, 1982). Fractals are generally self-similar on multiple scales. So, all fractals have a built-in form of iteration or recursion. Sometimes the recursion is visible in how the fractal is constructed. For example, Koch snowflake, Cantor set, and the Sierpinski triangle (figure 1a) are all generated using simple recursive rules. Self similarity is present in nature, too. Figures 1b shows a natural fractal object, the cauliflower which has the property of self similarity (it repeats its shape in different scales).

The self similarity and other kinds of fractals (e.g., Iterated Function Systems and the Lindenmayer System) are applied in different fields of computer science, for example, in computer graphics, in virtual reality, and in the computer networks (for controlling the traffic) (Leland et al, 1992; Erramilli et al., 1993; Liu, 2006).

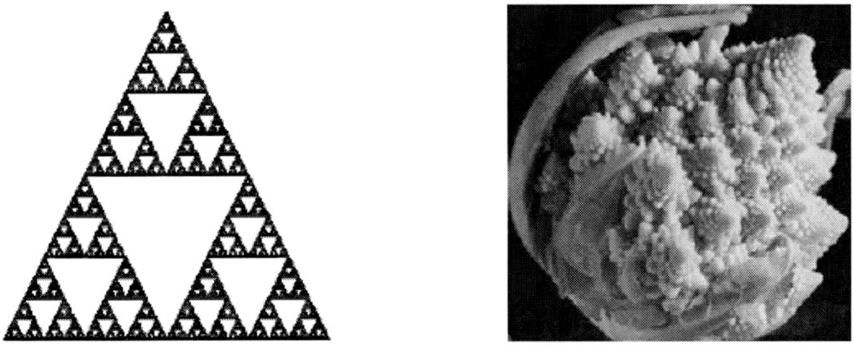

a) b)

Figure 1. The Sierpinski triangle is a fractal generated using simple geometric rules a). The cauliflower is a fractal present in the nature b).

2.1. The Self-Similarity

The invariance against changes in scale or size is named "self-similarity", and it is a property by which an object contains smaller copies of itself at arbitrary scales. Mandelbrot defined the self-similarity as follow: "When each piece of a shape is geometrically similar to the whole, both the shape and the cascade that generate it are called self-similar" (Mandelbrot, 1982, p. 34). "Similar" means that the relative proportions of the internal angles and shapes' sides remain the same. Mandelbrot (1982) observed that this property is ubiquitous in the natural world, and in the human body is possible to observe the presence of fractal geometry using two different points of view: temporal fractals and spatial fractals.

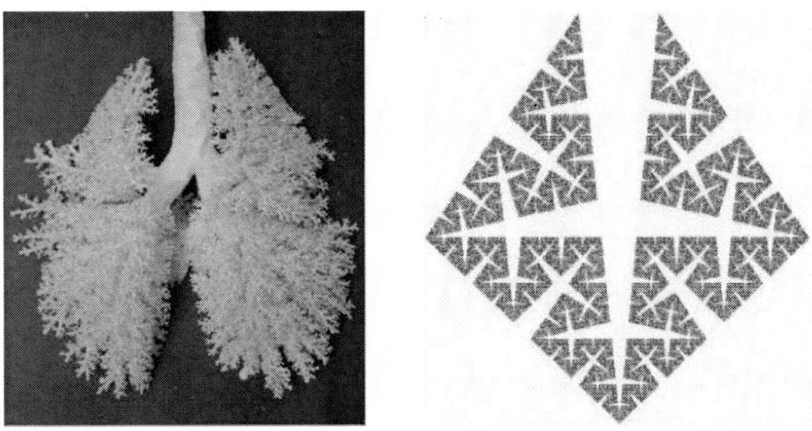

a) b)

Figure 2. Frontal view of dog's lungs a) it is a natural fractal object. "Fractal lungs" b) is a fractal generated using simple geometric rules, starting from two triangles.

Temporal fractals are present in some dynamic processes, for example in the cardiac rhythm. Spatial fractals refer to the presence of self-similarity observed to various enlargements, for instance, the small intestine repeats its form on different scales. Spatial fractals also refer to the branched patterns that are present inside the human body for enlarging the available surface for the absorption of the substances, and the distribution and the collection of the solutes occupying a relative small fraction of the body.

Figure 2a above shows the frontal view of dog's lungs. It is an example of self-similarity in the nature. In figures 2b there is an example of a fractal object which represents an attempt to reproduce complex shapes of the lungs, using few simple geometric rules.

2.2. The Iterated Function System

Iterated Function System (IFS) is another fractal that can be applied in computer science, in particular in computer graphics. Barnsley (1993, p. 80) defined the Iterated Function System as follow: "A (hyperbolic) iterated function system consists of a complete metric space (X, d) together with a finite set of contraction mappings w_n: X→ X with respective contractivity factor s_n, for n = 1, 2,.., N. The abbreviation "IFS" is used for "iterated function system". The notation for the IFS just announced is {X, w_n, n = 1, 2,.., N} and its contractivity factor is $s = \max\{s_n : n = 1, 2, \ldots, N\}$." Barnsley put the word "hyperbolic" in parentheses because it is sometimes dropped in practice. He also defined the following theorem (Barnsley, 1993, p. 81): "Let {X, w_n, n = 1, 2, ..., N} be a hyperbolic iterated function system with contractivity factor s. Then the transformation W: H(**X**) → H(**X**) defined by:

$$W(B) = \cup_{n=1}^{n} w_n(B) \qquad (1)$$

For all B∈ H(X), is a contraction mapping on the complete metric space (H(X), h(d)) with contractivity factor s. That is:

$$H(W(B), W(C)) \leq s \cdot h(B,C) \qquad (2)$$

for all B, C ∈ H(**X**). Its unique fixed point, A ∈ H(X), obeys

$$A = W(A) = \cup_{n=1}^{n} w_n(A) \qquad (3)$$

and is given by $A = \lim_{n\to\infty} W^{\circ n}(B)$ for any B ∈ H(X)."

The fixed point A ∈ H(X), described in the theorem by Barnsley is called the "attractor of the IFS" or "invariant set".

An affine map of the plane is given by form:

$$w\begin{bmatrix}x\\y\end{bmatrix}=\begin{bmatrix}a & c\\b & d\end{bmatrix}\begin{bmatrix}x\\y\end{bmatrix}+\begin{bmatrix}e\\f\end{bmatrix}$$

and it is determined by six number, *a*, *b*, *c*, *d*, *e*, and *f*. The affine maps are combinations of translations, rotations and scalings in the plane. If the scaling factor is less than 1, we have contractive affine maps.

Bogomolny (1998) affirms that two problems arise. One is to determine the fixed point of a given IFS, and it is solved by what is known as the "deterministic algorithm".

The second problem is the inverse of the first: for a given set A∈H(X), find an iterated function system that has A as its fixed point (Bogomolny, 1998). This is solved approximately by the Collage Theorem (Barnsley, 1993, p. 94).

The Collage Theorem states: "Let (X, d), be a complete metric space. Let L∈H(X) be given, and let $\varepsilon \geq o$ be given. Choose an IFS (or IFS with condensation) $\{X, (w_n), w_1, w_2, \ldots, w_n\}$ with contractivity factor $0 \leq s \leq 1$, so that

$$h(L, \cup_{\substack{n=1\\(n=0)}}^{n} w_n(L)) \leq \varepsilon \tag{4}$$

where h(d) is the Hausdorff metric. Then

$$h(L, A) \leq \frac{\varepsilon}{1-s} \tag{5}$$

where A is the attractor of the IFS. Equivalently

$$h(L, A) \leq (1-s)^{-1} h(L, \cup_{\substack{n-1\\(n-0)}} w_n(L)) \tag{6}$$

for all L∈H(X)."

The Collage Theorem describes how to find an Iterated Function System whose attractor is "close to" a given set, one must endeavour to find a set of transformations such that the union, or collage, of the images of the given set under transformations is near to the given set.

2.3. L-Systems

Hungarian biologist Aristid Lindenmayer (1925-1989) introduced a kind of fractals, called L-systems, for modelling biological growth in 1968. L-system or Lindenmayer system is an algorithmic method for generating branched forms and structures such as plants. The components of an L-system are the following:

- An alphabet which is a finite set V of formal symbols containing elements that can be replaced (variables).
- the constants which is a set S of symbols containing elements that remain fixed.
- The axiom (also called the initiator) which is a string ω of symbols from V defining the initial state of the system.
- A production (or rewriting rule) P that is a set of rules or productions defining the way variables can be replaced with combinations of constants and other variables. A production consists of two strings - the predecessor and the successor.

The rules of the L-system grammar are applied iteratively starting from the initial state.

L-systems are also commonly known as parametric L systems, and they are defined as a tuple $G = \{V, S, \omega, P\}$. L-system can be also defined as a formal grammar (a set of rules and symbols) most famously used for modelling the growth processes of plant development, and it has been thought able for modelling the morphology of a variety of organisms. The differences between L-systems and Chomsky grammars are well described by Prusinkiewicz and Lindenmayer that affirmed: "The essential difference between Chomsky grammars and L-systems lies in the method of applying productions. In Chomsky grammars productions are applied sequentially, whereas in L-systems they are applied in parallel and simultaneously replace all letters in a given word. This difference highlights the biological motivation of L-systems. Productions are intended to capture cell divisions in multicellular organisms, where many divisions may occur at the same time. Parallel production application has an essential impact on the formal properties of rewriting systems" (Prusinkiewicz and Lindenmayer, 1990, pp. 2 – 3). Strings generated by L-systems may be interpreted geometrically in different ways.

Table 1. Commands for LOGO-style turtle derived by L-systems

Symbols	Meaning
F	Move forward a step of length s. The state of the turtle changes, now it is (x', y', α), where $x' = x + s \cdot \cos\alpha$ and $y' = y + s \cdot \sin\alpha$. A segment between (x, y), starting point, and the point (x', y') is drawn.
f	Move forward a tep of length s without drawing a line.
+	Turn left by angle δ. The positive orientation of angles is counterclockwise, and the next state of the turtle is (x, y, α+δ).
-	Turn right by angle δ. The next state of the turtle is (x, y, α - δ).
[	Push the current state of the turtle onto a pushdown operations stack. The information saved on the stack contains the turtle's position and orientation, and possibly other attributes such as the color and width of lines being drawn.
]	Pop a state from the stack and make it the current state of the turtle. No line is drawn, although in general the position of the turtle changes.

For example, L-system strings serve a drawing commands for LOGO-style turtle. Prusinkiewicz and Lindenmayer defined a state of the turtle as a triplet (x, y, α), where the Cartesian coordinates (x, y) represent the turtle's position, and the angle α, called the heading, is interpreted as the direction in which the turtle is facing. Given the step size s and the angle increment δ, the turtle can respond to commands represented by the symbols in the table 1 above.

The Koch snowflake can be simply encoded as a Lindenmayer system with initial string "F++F++F", string rewriting rule F → F-F++F-F, and angle 60°, as shown in figure 3.

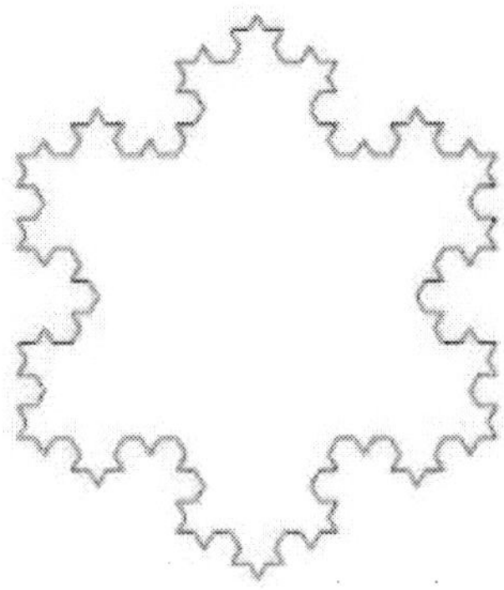

Figure 3. Koch snowflake generated by L-system.

3. Fractal Geometry in Computer Graphics

The terms fractals has been generalized by the computer graphics community and it includes objects outside Mandelbrot's original definition (Foley et al., 1997). It means anything that has a substantial measure of exact or statistical self-similarity. In the case of statistical fractals it is the probability density that repeats itself on every scale. An application field of fractal geometry is in the compression of images (fractal compression). A fractal compressed image can be defined as follow: it is an encoding that describes (i) the grid partitioning (the range blocks), and (ii) the affine transformations (one per range block) (Shulman, 2000). Research on fractal image compression derived from the mathematical ferment on chaos and fractals in the years 1978–1985. Barnsley was the principal researcher who worked in this field. The basic idea was to represent by an Iterated Function System (IFS) of which the fixed point is close to that image (Barnsley, 1988; Barnsley et al., 1988). This fixed point is also known as "fractal" (Fisher, 1995). Each IFS is coded as a contractive transformation with coefficients. The Banach fixed point theorem (1922), also known as the contraction mapping theorem or contraction mapping principle, guarantees the existence and uniqueness of fixed points of certain self maps of metric spaces, and provides a constructive method to find those fixed points.

An image can be represented using a set of IFS codes rather than pixels. In this way, a good compression ratio could be achieved. This method was good for generating almost real images based on simple iterative algorithms. The inverse problem, going from a given image to an Iterated Function System that can generate the original (or at least closely resemble it), was solved by Jacquin according to Barnsley in March 1988. He introduced a modified scheme for representing images called Partitioned Iterated Function Systems (PIFS). The main characteristics of this approach were that (i) it relied on the assumption that image redundancy can be efficiently exploited through self-transformability on a block-wise basis, and (ii) it approximated an original image by a fractal image (Jacquin, 1992). In a PIFS, the transformations do not map from the whole image to the parts, but from larger parts to smaller parts. In the Jacquin's method, the small areas are called "range blocks", the big areas are called "domain blocks" and the pattern of range blocks was called the "partitioning of an image". Every pixel of the original image has to belong to one range block. This system of mappings is contractive, thus when iterated it quickly converge to its fixed point image. Thus, the key point for this algorithm is to find fractals which can best describe the original image and then to represent them as affine transformations.

All methods are based on the fractal transform using iterated function systems which generate a close approximation of a scene using only few transformations (Peitgen and Saupe, 1988; Wohlberg and de Jager, 1999; Zhao and Liu, 2005).

Fractal compression is a lossy compression method (compressing data and then decompressing it retrieves data that may well be different from the original, but is close enough to be useful in some way), and most lossy compression formats suffer from generation loss: repeatedly compressing and decompressing the file will cause it to progressively lose quality (as shown in figure 4).

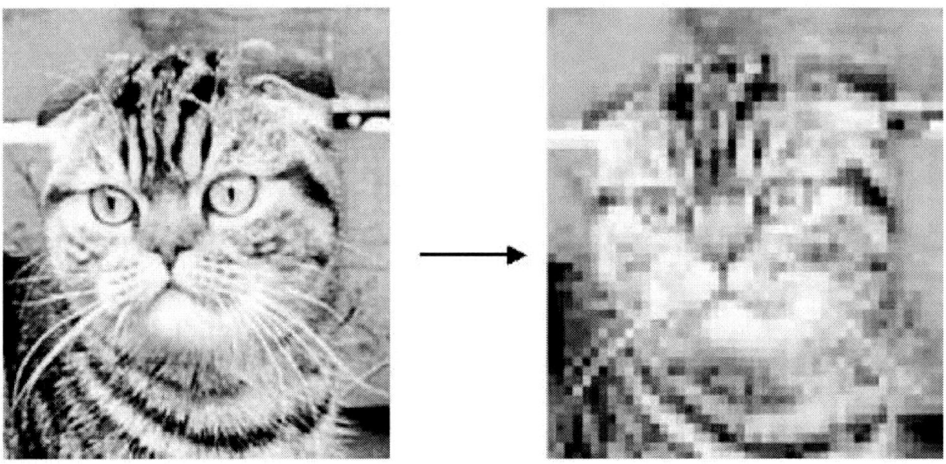

Figure 4. Fractal compression: repeatedly compressing and decompressing the file will cause it to progressively lose quality.

In the field of the compression of images novel methods are studied. For example, van Wijk and Saupe (2004) present a fast method to generate fractal imagery based on Iterated Function Systems. The high performance derives from three factors: (i) all steps are expressed in graphics operations which can be performed quickly by hardware, (ii) frame to frame coherence is exploited, (iii) only a few commands per image have to be passed from the CPU to the graphics hardware, avoiding the classical CPU bottleneck.

4. Fractal Geometry for Modelling Virtual Landscapes

Other interesting application of fractal geometry in computer science is for modelling landscapes which include terrain, mountains, trees and so forth. Fournier et al. (1982) developed a mechanism for generating a kind of fractal mountains based on recursive subdivision algorithm for a triangle. Here, the midpoints of each side of the triangle are connected, creating four new subtriangles. Figure 5a shows the subdivision of the triangle into four smaller triangle, figure 5b illustrates how the midpoints of the original triangle are perturbed in the y direction (Foley et al., 1997). To perturb these points, can be use the properties of the self-similarity, and the conditional expectation properties of fractional Brownian motion (abbreviated to fBm). The fractional Brownian motion was originally introduced by Mandelbrot and Van Ness in 1968 as a generalization of the Brownian motion (Bm). FBm basically consists of steps in a random direction and with a step-length that has some characteristic value. Hence the random walk process. An important feature of fBm is the self-similarity (if we zoom in on any part of the function we will produce a similar random walk in the zoomed in part). Figure 6 shows a recursive subdivision of an initial polygon using squares. Other polygons can be used to generate the grid (e.g., triangles and hexagons).

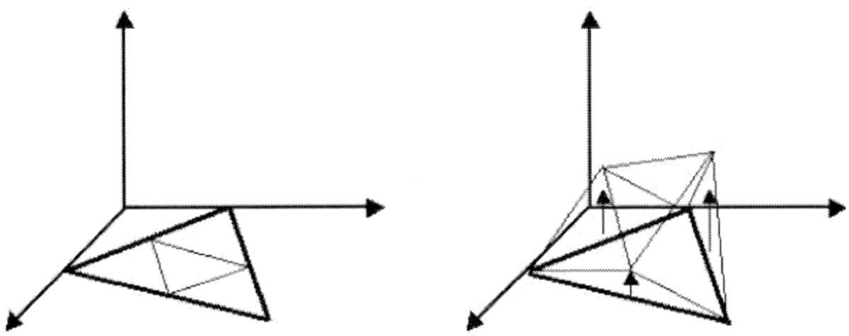

a) b)

Figure 5. The subdivision of a triangle into four smaller triangle a). Perturbation in the y direction of the midpoints of the original triangle b).

This method evidences two problems, which are classified as internal and external consistency problems (Fournier et al., 1982, pp. 374-375). Internal consistency is the reproducibility of the primitive at any position in an appropriate coordinate space and at any level of detail, so the final shape is independent of the orientation of the subdivided triangle. This is satisfied by a Gaussian random number generator which depends on the point's position, thus it generates the same numbers in the same order at a given subdivision level. The external consistency concerns the midpoint displacement at shared edges and their direction of displacement. This process, when iterated, produces a deformed grid which represents a surface, an example is shown in figure 6. After the rendering phase (that includes: hidden line, coloured, and shaded) can appear a realistic fractal mountain, as shown in figure 7.

These examples describe how to realize fractal mountains but not their erosion. Musgrave et al.. (1989) introduced techniques which are independent of the terrain creation. The algorithm can be applied to already generated data represented as regular height fields require separate processes to define the mountain and the river system. Prusinkiewicz and Hammel (1993) combined the midpoint-displacement method for mountain generation with the squig-curve model of a non-branching river originated by Mandelbrot (Mandelbrot, 1978; 1979).

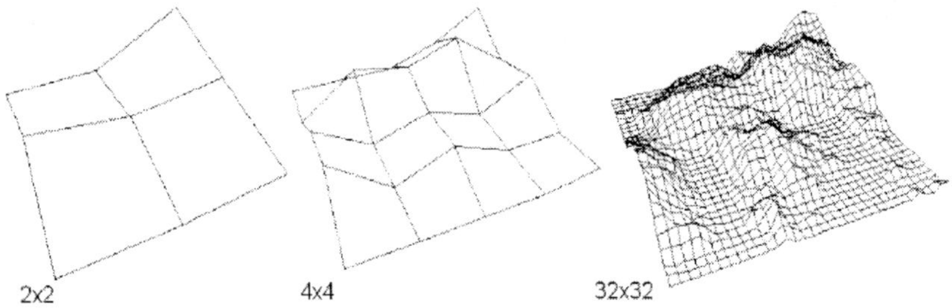

Figure 6. Grid of squares generated by a recursive subdivision and applying the fractional Brownian motion.

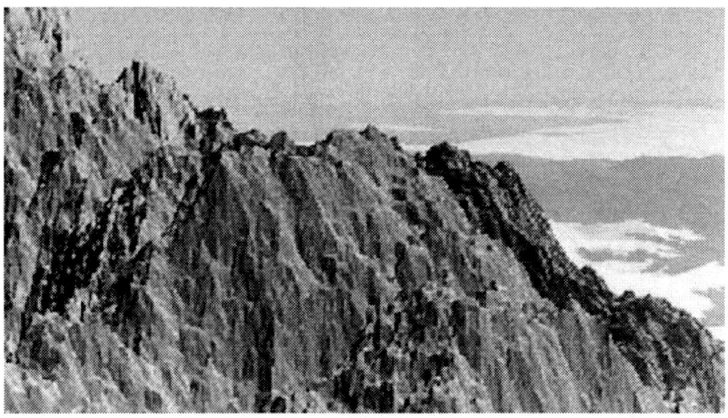

Figure 7. Fractal mountains.

Their method created one non-branching river as result of context sensitive L-system operating on geometric objects (a set of triangles). Three key problems remained open (i) the river flowed at a constant altitude, (ii) the river flowed in an asymmetric valley, and (iii) the river had no tributaries. Figure 8 shows an example of a squig-curve construction (recursion levels 0–7) (Prusinkiewicz and Hammel, 1993, p. 177).

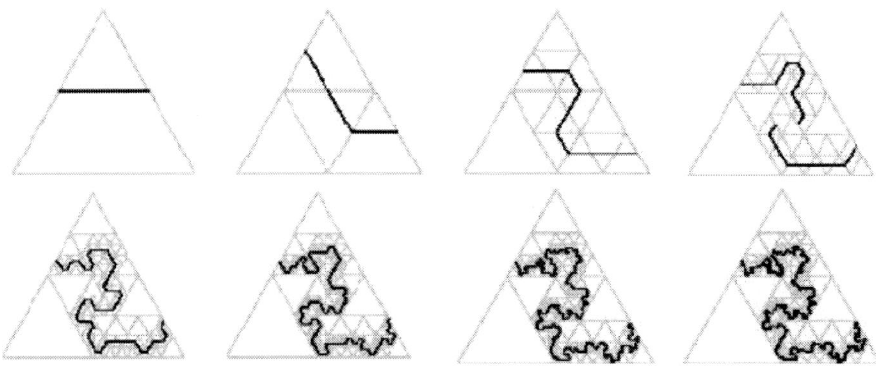

Figure 8. Squid-curve construction (recursion level 0-7).

Maràk et al. (1997) reported a method for synthetical terrain erosion, that is based on rewriting process of matrices representing terrain parts. They found a rewriting process which was context sensitive. It was defined as a set of productions A → B, where A and B were matrices of numbers of type: N ×N, where N>0.

The terrain parts were rewritten using certain user-defined set of rules which represented an erosion process. The method consisted in three kinds of rewriting process (Maràk et al., 1997; Maràk, 1997). The "absolute rewriting" which permitted to erode the objects in predefined altitude.

The "rewriting with the reference point" which could erode arbitrary object in any altitude. The third kind of rewriting process which permitted to erode some shapes in any scale. Its advantage was that the algorithm could be controlled by some external rules and it could simulate different kinds of erosion (Maràk et al., 1997).

Fractal algorithms can also be used in computer graphics to generate complex objects using Iterated Function Systems. Figure 9 shows a fern leaf created using the IFS. The IFS is produced by polygons, in this case: squares, that are put in one another. The final step of this iterative process shows a fern which has high degree of similarity to real one.

Originally the L-systems were devised to provide a formal description of the development of such simple multicellular organisms, and to illustrate the neighbourhood relationships between plant cells. Later on, this system was extended to describe higher plants and complex branching structures. Smith (1984) was the first to prove that L-systems were useful in computer graphics for describing the structure of certain plants, in his paper: "Plants, Fractals, and Formal Languages". He described that these objects should not be labeled as "fractals" for their similarity to fractals, introducing a new class of objects which Smith called "graftals".

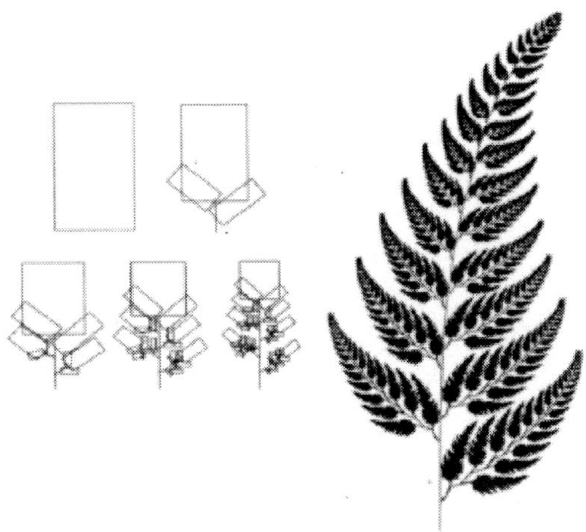

Figure 9. Fern leaf created using the IFS.

This class had of great interest in the Computer Imagery (Smith, 1984; Foley et al., 1997). Figure 10 shows an example of plant-like structures generated after four iterations by bracketed L-systems with the initial string F (angle δ= 22.5°), and the replacement rule F →FF+[+F-F-F] -[-F+F+F].

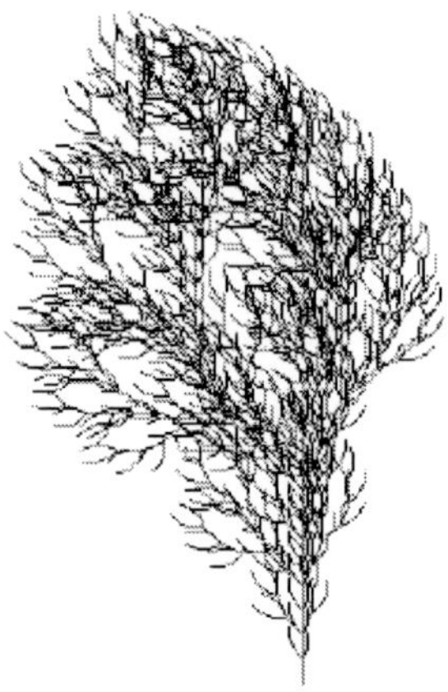

Figure 10. Plant-like structures generated after four iterations by bracketed L-systems.

Using fractal algorithms is possible to create virtual mountains and virtual trees described in Virtual Reality Modelling Language (VRML), as shown in figure 11a. VRML is a 3D graphics language used on the World Wide Web for producing "virtual worlds" that appear on the display screens using an appropriate VRML browser. This example shows that the connections between fractal geometry, and the "virtual worlds" exit. A virtual world is a computer-based simulated environment intended for its users to inhabit and interact via avatars. Avatars can walk on virtual territory generated with fractal algorithms. Figure 11b shows a fractal tree generated with simple geometric iterative rules.

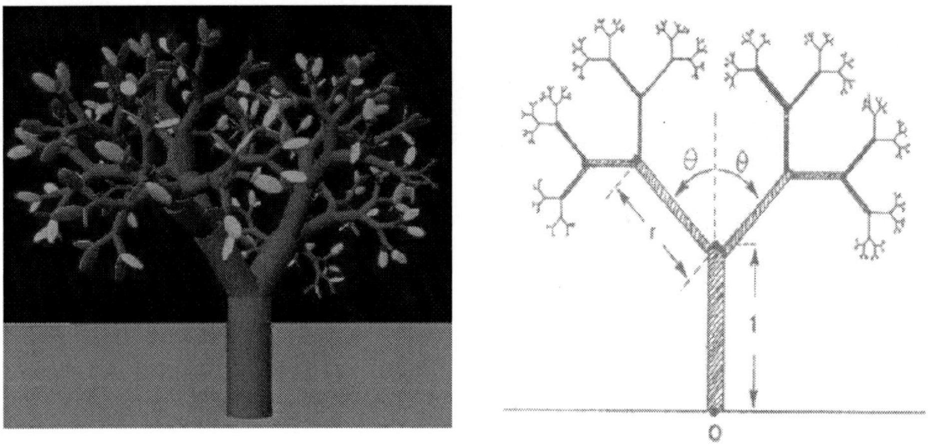

a) b)

Figure 11. Virtual tree realized in VRML using fractal algorithms a). A fractl geometric tree b).

The evolution of the computer graphics techniques and virtual reality technologies, also connected to the fractal algorithms, help to create a new kind of digital arts. Some artists represent a broad range of approaches to installation and virtual reality art (Wands, 2006). For example, Char Davies is a pioneer of this art form. She describes her immersive virtual reality environments *Osmose* (1995) and *Ephémère* (1998) as works "known for their embodying interface, painterly aesthetic and evocation of landscape".

Figure 12 shows the installation space for Osmose. Davies and Harrison claim: "*Osmose* is an immersive virtual environment, produced by Softimage in 1994/95. One of the primary goals of *Osmose* was to push the expressive capabilities of existing 3D tools, to demonstrate that an alternative aesthetic and interactive sensibility is possible for real-time, interactive, 3D computer graphics. *Osmose* was created under the direction of Char Davies, the Director of Visual Research at Softimage.(...) One of Davies' intentions for Osmose was to create a space that is "psychically innovating," one in which, to quote Bachelard, participants do not change "place," but change their own nature. Osmose was therefore designed to explore the potential of immersive virtual space to allow participants to shed their habitual ways of looking at (and behaving in) the world." (Davies and Harrison, p. 25, 1996).

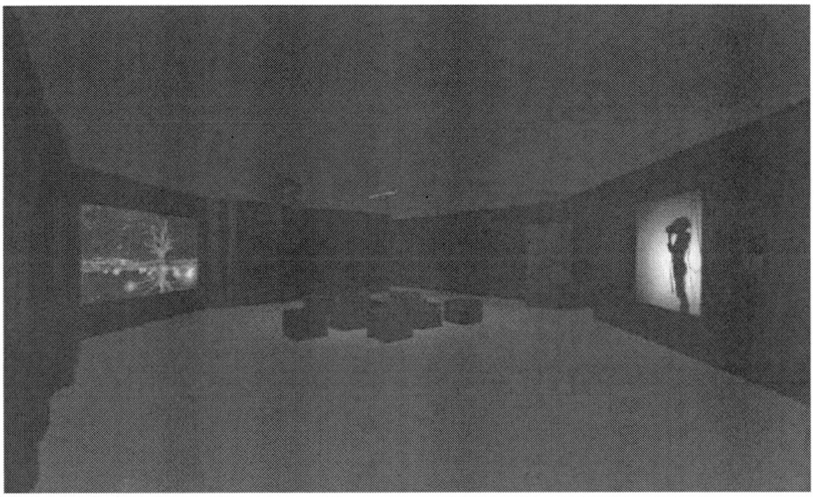

Figure 12. The installation space for Osmose (1995). Char Davies/Immersence Inc.and Softimage Inc.©.

CONCLUSIONS

This paper has described some applications of fractal geometry in computer graphics. The self-similarity, important characteristic of the fractal objects, is a unifying concept, and it is an attribute of many laws of nature and innumerable phenomena in the world around us.

In computer graphics, fractals can be applied in different fields: to compress the images using simple algorithms based on Iterated Function Systems, for modelling complex objects in computer graphics (e.g., mountains, trees and rivers) using L-systems and the fractional Brownian motion. Future trends are oriented to use IFS to generate terrain from real data extracted from geological data base. This is useful in the reconstruction of real terrain and landscapes (Guérin et al., 2002; Guérin, and Tosan, 2005). Fractal geometry also offers an alternative approach to conventional planetary terrain analysis. For example, Stepinski et al. (2004) describe Martian terrains, represented by topography based on the Mars Orbiter Laser Altimetry (MOLA) data, as a series of drainage basins. Fractal analysis of each drainage network, computationally extracts some network descriptors that are used for a quantitative characterization and classification of Martian surfaces.

The Computer graphics connected to the fractal geometry and to the virtual reality have also given to the digital artists powerful tools and new sources of creative expression (davies and Harrison, 1996; Thwaites, 2005; Wands, 2006).

REFERENCES

Barnsley, M.F. (1988). *Fractals everywhere*. Boston: Academic Press.

Barnsley, M.F. (1993). *Fractals everywhere*. Boston: Academic Press, 2nd edition.

Barnsley, M.F., Saupe, D., and Vrscay, E.R. (Eds.) (2002). *Fractals in Multimedia*. Berlin, Germany: Springer.

Barnsley, M.F., Jacquin, A.E., Malassenet, F., Reuter, L., and Sloane, A.D. (1988). Harnessing chaos for image synthesis. *SIGGRAPH 1988*, pp. 131-140.

Bogomolny, A.(1998). *The Collage Theorem*. Retrieved September 1, 2006, from: http://www.cut-the-knot.org/ctk/ifs.shtml.

Davies, C., and Harrison, J. (1996). Osmose: Towards Broadening the Aesthetics of Virtual Reality. Computer Graphics (ACM SIGGRAPH), Vol. XXX (4) (1996) pp. 25-28.

Eglash, R. (1999). *African Fractals: Modern Computing and Indigenous Design*. Piscataway, NJ: Rutgers University Press.

Erramilli, A., Gosby, D., and Willinger, W. (1993). Engineering for Realistic Traffic: A Fractal Analysis of Burstiness. *Proceedings of ITC Special Congress*, Bangalore, India, 1993.

Fisher, Y. (1995). *Fractal Image Compression: Theory and Application*. New York: Springer-Verlag.

Foley, J.D., van Dam, A., Feiner, S.K., and Hughes, J.F. (1997). *Computer Graphics: Principles and Practice. Second Edition in C*. New York: Addison Wesley.

Fournier, A., Fussel, D., and Carpenter, L. (1982). Computer Rendering of Stochastic Models. *Communications of the ACM*, 25, pp. 371-384.

Fractal geometry. (2007). In *Encyclopædia Britannica*. Retrieved February 2, 2007, from Britannica Concise Encyclopedia: http://concise.britannica.com/ebc/article-9364797/fractal-geometry.

Guérin, E., and Tosan, E. (2005). Fractal Inverse Problem: Approximation Formulation and Differential Methods. Lévy-Véhel, J., and Lutton E. (Eds.). *Fractal in Engineering: new trends in theory and applications* (pp. 271-285). London: Springer.

Guérin, E., Tosan, E., and Baskurt, A. (2002). Modeling and Approximation of Fractal Surfaces with Projected IFS Attractors". Novak, M.M. (Ed.). *Emergent Nature: Patterns, Growth and Scaling in the Science* (pp. 293 – 303). New Jersey: World Scientific.

Leland,W.E., Taqqu, M.S., Willinger, W., and Wilson, D.V. (1993). On the Self-Similar Nature of Ethernet Traffic. *Proceedings of the ACM/SIGCOMM'93*, (pp. 183-193) San Francisco, CA.

Mandelbrot, B. (1975). *Les Objects Fractals. Forme, Hasard et Dimension,* Paris, France : Nouvelle Bibliothèque Scientifique Flammaron.

Mandelbrot, B. (1982).*The Fractal Geometry of Nature*. W.H. Freeman and Company.

Maràk I. (1997). On Synthetic Terrain Erosion Modeling: A Survey. Retrieved March 14, 2007, from. http://www.cescg.org/CESCG97/marak/

Marák, I., Benes, B., and Slavík, P. (1997) Terrain Erosion Model Based on Rewriting of Matrices. *Proceedings of WSCG-97*, vol. 2., (pp. 341-351).

Musgrave, F.K., Kolb, C.E., and Mace, R.S. (1989). The synthesis and rendering of eroded fractal terrain. *Proceedings of SIGGRAPH '89*, in *Computer Graphics* 23, 3, (pp.41–50), ACM SIGGRAPH, New York.

Nonnenmacher, T.F., Losa, G.A., Merlini, D., and Weibel, E.R. (Eds.). (1994). *Fractal in Biology and Medicine*. Basel, Switzerland: Birkhauser.

Peitgen, H., and Saupe, D. (1988). *The Science of Fractal Images*. New York: Springer-Verlag.

Prusinkiewicz, P. and Hammel, M. (1993). A Fractal Model of Mountains with Rivers. *Proceeding of Graphics Interface '93*, pp. 174-180.

Prusinkiewicz, P., and Lindenmayer, A. (1990). *The Algorithmic Beauty of Plants*. New York, US: Springer-Verlag. Retrieved September 10, 2007, from:
http://algorithmicbotany.org/papers/abop/abop.pdf

Sala, N. (2006). Complexity, Fractals, Nature and Industrial Design: Some Connections, Novak, M.M. (Ed.). *Complexus Mundi: Emergent Pattern in Nature*. (pp. 171 – 180). Singapore: World Scientific.

Stepinski, T.F., Collier, M.L., McGovern, P.J., and Clifford, S.M. (2004). Martian geomorphology from fractal analysis of drainage networks. *Journal of Geophysical Research*, Vol. 109, n°E2, pp. E02005.1-E02005.12.

Thwaites, H. (2005). The Immersant Experience of Osmose and Ephémère." [PDF File] Proceedings of the 2005 international conference on Augmented tele-existence ICAT '05. Christchurch, New Zealand, pp. 148–155.

van Wijk J. J., andSaupe D. (2004). Image based rendering of iterated function systems. *Computers and Graphics*, 28(6), 937-943.

Vyzantiadou, M.A., Avdelas, A.V., and Zafiropoulos, S.(2007). The application of fractal geometry to the design of grid or reticulated shell structures. *Computer-Aided Design*, Volume 39, issue 1, 51-59.

Wands, B. (2006). Art of the Digital Age. New York, NY: Thames and Hudson Inc.

Wohlberg, B., de Jager, G. (1999). A Review of the Fractal Image Coding Literature. *IEEE Transactions on Image Processing*, 8(12), 1716-1729.

Zhao, E., and Liu, D. (2005). Fractal image compression methods: a review. *ICITA 2005.Third International Conference on Information Technology and Applications*, Volume 1 (pp. 756 – 759).

In: Progress in Chaos Complexity Research
Editors: Franco F. Orsucci and Nicoletta Sala
ISBN: 978-1-60456-375-7

Chapter 11

BUYER DECISIONS IN THE US HOUSING INDUSTRY

Michael Nwogugu*
Certified Public Accountant, Maryland, USA
City College Of New York. Columbia University,
Attended Suffolk University Law School, Boston, USA

ABSTRACT

This article: 1) develops new psychological theories and mathematical models that can explain many of the legal and economic problems that occurred in the US housing industry between 2000 and the present – such as the sub-prime loan problems, predatory lending, mortgage fraud, title defects, rapid/un-warranted price increases and sales fraud; 2) analyzes and identifies the psychological and behavioral biases of first-time homebuyers and repeat home buyers, 3) develop new theories (testable hypothesis) of psychological effects and Biases inherent in the housing purchase/sale process; 4) develops theoretical mathematical models for Buyers' Propensity-To-Purchase.

This study involves analysis of historical economic trends, critique of existing methods and theories, and development of new theories and development of mathematical models. This article also partly relies on surveys and published empirical research using US macroeconomic and housing data from the 1995-2003 period. At the present time, the models developed in this article cannot be realistically tested empirically because real estate data and price series and psychological effects described in the models (and associated periodic changes in such data or the logarithms of such data) don't fit known distributions and regression techniques.

Keywords: *Housing, urban economics, decision analysis, risk, complexity, macro-economics.*

* Address: P. O. Box 170002, Brooklyn, NY 11217, USA. Phone/Fax: 1-718-638-6270. Email: datagh@peoplepc.com, mcn111@juno.com

INTRODUCTION

Psychological factors played an important role in many trends and changes in the US housing industry between 1995 and 2003. Much of these issues have not been analyzed theoretically or empirically in the existing literature. Garvill, Garling and Lindberg and Montgomery (1992); Zietz (2003); Gunnelin, Hendershott and Hoesli (2004); Militino, Ugarte and Garcia-Reinaldos (2004); Kauko (2004); Leung, Lau and Leong (2002); Haurin, Parcel and Haurin (2002); Boyle and Kiel (2001); Watkins (1998). Himmelberg, Mayer and Sinia (2005).

Housing accounts for more than 80% of the built environment in most countries, and for a substantial portion of household wealth in the US. Between 2000-2005, the total value of residential property in developed economies increased by more than $30 trillion, to over $70 trillion, an increase equivalent to 100% of those countries' combined GDPs. The apparently global boom in house prices during 2000-2005 was attributable to historically low interest rates which encouraged home buyers to borrow more money; and loss of faith in stock markets after they declined in value substantially and thus, made real estate look attractive. (Source: *The Economist*, June 2005). Muller and Riedl (2002).

The housing industry/market contributed to propping up the US economy during 1995-2003, and a sharp decline in housing prices is likely to have significant negative consequences. Between 2000-2005, US consumer spending and residential construction accounted for 90% of the total growth in US GDP. More than two-fifths of all private-sector jobs created in the US between 2001-2005 were in housing-related sectors, such as construction, real estate and mortgage brokerage. Its clear that the true state of the US economy was, and continued to be masked by significant borrowing used to finance construction, property acquisitions, and consumer spending. Without such credit availability, the US economy, productivity, GDP and GDP growth would have declined during 1995-2003. This illusion is in contrast to the economies of many countries where credit is much more scarce.

BUYERS' PSYCHOLOGICAL BIASES AND EFFECTS IN THE US HOUSING MARKETS

In the US, psychological factors apparently played a major role in peoples'/households' Propensity-To-Purchase between 1995-2003. Malpezzi (1996); Rosen (1979); Henderson and Ionnides (1989); Ihlanfeldt (1980); Linneman (1986); Vigdor (December 2004). Akhter (2003). Olsen, Donaldson and Pereira (2003). Antonides and Kroft (2005); Butler (2000). Ross and Sohl (2005). Berkowitz and Hynes (1999). Pingle and Mitchell (2002). Stevenson (2004). Gallimore (2004). The following are key hypothesis about psychological effects and biases associated with seller's propensity-to-sell and buyers' propensity-to-purchase.

Proposition 1 - Wealth Bias – people's propensity to buy housing units depends on their perceptions of wealth and utility gained from holding or controlling wealth in specific forms. Real estate represents a form of wealth that is more immediate and tangible, and hence provides more satisfaction in consumption than many other forms of wealth. Benjamin, Chinloy and Jud (2004, 2002); Benito and Wood (2005); Levy and Lee (2004); Iwarere and

Williams (2003); Andrew and Meen (2003); Boelhouwer (2002); Donkers and Van Soest (1999); Green (2002); Raess and Von Ungern-Sternberg (2002); Aizecorbe, Kennickell and Moore (2003); Meier, Kirchler and Hubert (1999); Krystalogianni and Tsocalos (2004). Home-equity is a form of wealth that is created subtly with one of the least "external resistance" or 'transactional/interactional friction' from the environment. Some forms of wealth may be perceived differently than others, and thus, may have higher/lower values to different people. This *Wealth Bias* is different from Framing-Effects because:

- For any given housing unit, the range of available and relevant knowledge is relatively constant and stable.
- Framing requires changes in major characteristics of the scenario or object or situation, which is not the case in these instances.

Income and capital appreciation from residential property may cause a Wealth Bias in which the Buyer's Propensity-to-Buy will vary depending on his/her classification, perception, gained-utility, use-value and or possessory-value of the historical income, current income, capital appreciation and expected income from the property; and his/her preferences, tax position, Willingness-To-Accept-Losses ("WTAL"), loss aversion, aspirations, family structure, opportunity costs, etc.. Hence, the Buyer's perception of home equity and or income from real estate may create incentives or disincentives to buy a property. Perceptions of net-wealth versus total wealth is a relevant psychological effect and influences the decision to buy or sell a housing unit. Case, Quigley and Shiller (2005); Malpezzi, Englund, Kim and Turner (2004); Boelhouwer (2002); Hurst and Stafford (2005); Cocco and Campbell (2003); Bennett and Peach and Stavros (1998). The leverage used in most property purchases creates a perception of increased wealth, in which the buyer is unconsciously fixated on the total value of the property and not the net value (net of debt, transaction costs, maintainance costs and taxes). This is in contrast to other common forms of investment such as stocks and bonds where margin is not used for all purchases (margin is typically 50%), and the difference between net wealth and total wealth is much smaller. Furthermore, utility from wealth may be more realistic if the buyer intends to occupy the target property.

Proposition 2 - Tenure Bias: People's perceptions of rental-tenure versus ownership-tenure may have also contributed to changes in home prices and housing demand in the US during 1995-2003 – this is referred to as Tenure Differential. Malpezzi (1996); Malpezzi, Englund, Kim and Turner (2004); Henderson and Ionnides (1989); Ihlanfeldt (1980); Linneman (1986); Raess and Von Ungern-Sternberg (2002); Bennett, Peach and Peristiani (1998). Tenure Bias affects Buyer propensity-to-Buy, because it shapes Buyers' expectations about capital commitments, investment horizons, ownership rights, and expected returns. Tenure Bias has different effects on: 1) renters who are trying to decide whether or not to buy or rent, and 2) home owners who want to decide whether to sell and then rent or buy another property. Tenure Bias is conjectured to vary among households depending on the age of the head of household, education, knowledge, wealth and preferences. In many instances, people tend to view rental tenure as more of a shorter term arrangement, involving less emotions and financial commitment to the housing unit and also lower investment returns; and conversely, view ownership tenure as a long term arrangement with greater monetary, non-monetary and emotional commitments and greater investment returns. Such perceptions are either a major consideration in investment decisions or are incorporated into the individual's selection of an investment horizon, and are shaped by the neighborhood, type of landlord, type of housing unit, expectations about increases in home values, household income, expected future financial commitments, knowledge, family structure, etc.. In the US, the Tenure Differential was manifested during 1995-2003 by: 1) the annual changes in the percentage of renters who choose to purchase homes; 2) the annual changes in the percentage of homeowners who choose to sell homes, and then rent housing units.

Proposition #3: Financing Effect - The form of financing of home purchases also has psychological implications:

- The Duration of pre-financing processes (pre-application, qualification, application, notification; online vs. physical procedures, etc.,), affects the amount of stress, and the amount of sub-consciously-remembered stress and friction associated with the transaction.
- Effect of property valuation.
- Psychological impact of approval process (different financing methods have different approval processes which have different levels of disclosure, acceptance/rejection rates, and hence different psychological effects).
- Post-transaction psychological adjustments (worries about credit scores, budgeting, job stress, income allocation, evaluation of opportunity costs, regret, etc.).

See: Bennett, Peach and Peristiani (1998). The use of debt financing requires less capital outlay, and results in more tax benefits, and this has substantial psychological implications for buyers:

- Mortgages reduce perceived total risk to the buyer.
- Mortgages provide better ability to face future uncertainty because the buyer has more cash compared to all-equity financing.
- Mortgages provide the buyer with an incentive to hold a job, and to maintain a certain level of income.
- Mortgages cause increases in perceived wealth – the sub-conscious and or conscious psychological value of the property is often the total value of the property instead of the net-realizable-value (total value minus outstanding debt, transaction costs and maintainance costs).

Proposition 4: Comparison Effects – As part of their conscious and sub-conscious decision processes and search processes, prospective Buyers tend to compare the annual cost of ownership to the annual cost of renting a housing unit. Such comparisons are often done in various stages/phases over time, often with household members, and within the context of expected tenure, available downpayment, financability (credit record, income, etc.), expected returns, need for income and expected family structure. Hence, the Buyer's Propensity-To-Buy is affected by the annual cost of ownership. Bennett, Peach and Peristiani (1998); Raess and Von Ungern-Sternberg (2002); Himmelberg, Mayer and Sinai (2005). The widely accepted formula for the annual cost of ownership is explained in Himmelberg, Mayer and Sinai (2005), and is described as follows:

ϕ = foregone interest that the owner could have earned by investing an amount equal to the home price, in other investments. $\phi = (P_t * r_t)$

P_t = price of housing

r_t = risk free interest rate

ω_t = property tax rate

τ_t = income tax rate

π = annual Property taxes = $(P_t * \omega_t)$

Ψ = tax shield from deduction of mortgage interest payments and property tax payments= $(P_t \tau_t) * (r_t + \omega_t)$

δ_t = annual maintainance costs expressed as a percentage of of the price of the housing unit

g_{t+1} = expected capital gain or loss during the year.

λ_t = a risk premium that incorporates the higher risk inherent in ownership as opposed to renting. $\lambda_t = P_t \gamma_t$.

α = The annual cost of ownership = $\phi + \pi - \Psi + (P_t \delta_t) - (P_t g_{t+1}) + \lambda_t$

Himmelberg, Mayer and Sinai (2005), stated that the imputed rental value of housing is:

$$R_{it} = P_{it} - [\{(1 - \delta_{It})/(1 + r_t + \gamma_{it})\} * \{E_{it} * P_{it+1}\}] + \{(1 - \tau_{it})\omega i_t P_{it}\} - (\tau_t R_t P_{it})$$

Hence, if the annual Cost-of-ownership exceeds the annual cost-of-renting, then housing prices are likely to decline.

$$R_t = P_t \beta$$

where:

R_t = annual Cost of Renting

$\beta = r_t + \omega_t - \tau_t(r_t + \omega_t) + \delta_t - g_{t+1} + \gamma_t$.

However, the above-mentioned formulas are incorrect, primarily because they don't incorporate the value of time, several opportunity costs and psychological effects. For a given housing unit, and for a household that has to choose between rental and ownership of said housing unit, the following are the formulas for the cost-of-ownership and the cost-of-renting.

α = The annual cost of ownership =

$$\phi + I_{tt} + It_l + \varsigma_{at} + (P_t * \omega_t) - \Psi + \delta_{ot} - (E_t * P_t) - (P_t * g_{t+1}) + \lambda_t + (r_{dt} * P_t) + (P_{vt} * F_{vt}) + (P_{pt} * F_{pt}) + U_{tt} + U_{st} + S_{ot} + \theta$$

The annual cost of rental = $\varphi = R_t + \boldsymbol{\theta} + \delta_{rt} + (\mathbf{r_{rt}} * P_t) + (\mathbf{S_{rt}}) - \boldsymbol{\phi_t} + (\mathbf{P_t} * g_{t+1})$

where:

R_t = annual rent.

ϕ_t = opportunity cost; the foregone interest that the owner could have earned by investing an amount equal to the downpayment plus closing costs, in other investments. $\phi = (\varsigma_t * r_{bt})$

I_{lt} = property insurance costs – hazard, fire, arson, etc.

I_{tt} = title insurance costs

E_t = equity built up in year *t*, by payment of mortgage principal, if any; expressed as a percentage of the price of the house.

ς = cash down-payment.

ς_{at} = annual amortization of downpayment and closing costs – amortized over the average ownership tenure. r_{dt} = the difference between the borrower's overall cost of capital, 1) if debt is used to purchase the property versus 2) if equity only is used. Mortgages increase the borrower's post-transaction cost of capital for any other new/marginal borrowing.

r_{rt} = the difference between the renter's overall cost of capital, if debt is used to purchase the property versus if property is rented. Mortgages increase the renter/borrower's cost of capital for any other new/marginal borrowing.

S_{ot} = annual amortization of search costs for home purchase – amortized over the average ownership tenure.

S_{rt} = annual amortization of search costs for home rental – amortized over the average rental tenure.

r_i = annual mortgage interest rate (loan constant).

χ = annual mortgage interest payment

P_t = price of housing

r_{bt} = the annual yield on BBB-A rated corporate bonds with the remaining term equal to the average home ownership tenure.

r_t = risk free interest rate

ω_t = property tax rate

= annual depreciation in time t; expressed as a percentage of the price of the home.

τ_t = income tax rate

π = annual Property taxes = $(P_t * \omega_t)$

Ψ = tax shield from deduction of mortgage interest payments, depreciation and property tax payments = $(P_t \tau_t)*(r_i + \omega_t)$

δ_{ot} = annual maintainance cost, if property is owned. Includes value of owner's time, cost of materials, cost of hired labor, etc.

δ_{rt} = annual maintainance cost if property is rented. Includes value of tenant's time, cost of materials, cost of hired labor, etc.

Ω_t = the "Occupancy Effort" refers to the occupant's vigilance in maintaining the property. In many instances, the landlord is not responsible for some maintainance. Obviously, the owner-occupant will exert a higher Occupancy-effort than the tenant. $\delta_{rt}, \delta_{ot} \in \Omega_t$

Ω_d = the "Occupancy Effort Differential" refers to the difference between δ_{rt} and δ_{ot}, and refers to the vigilance in maintaining the property.

g_{t+1} = expected capital gain or loss during the year, expressed as a percentage of the price of the housing unit. $g_{t+1} \in (-\infty, \infty)$.

λ_t = a risk premium that incorporates the higher risk inherent in ownership as opposed to renting. $\lambda_t = P_t \gamma_t$.

U_{tt} = value of homeowner's time spent on house-related matters in time *t*.

U_{st} = value of reduced stress, if any; arising from homeownership in time *t*.

θ = cost of utilities in time t (electricity, HVAC, water).

P_{vt} = probability of occurrence of building code or local ordinance violations in time t.

F_{vt} = fine for violation of building code or local ordinance violation in time t.

P_{pt} = probability of occurence of event that causes premises liability and successful prosecution/lawsuit.

F_{pt} = fine/damages for premises liability.

The annual 'Indifference Rent' (R_{it}) at which the household is indifferent between ownership and rental, is calculated as follows:

$$R_{it} = \phi + I_{tt} + It_l + \varsigma_{at} + (P_t * \omega_t) - \Psi + (\delta_{ot}) - (E_t * P_t) - (P_t * g_{t+1}) + \lambda_t + (r_{dt} * P_t) + (P_{vt} * F_{vt}) + (P_{pt} * F_{pt}) + U_{tt} + U_{st} + S_{ot} + \theta$$

$$- \theta - r_{rt} - (S_{rt}) + \phi - (P_t * g_{t+1}) - \delta_{rt}$$

$$R_{it} = 2\phi + I_{tt} + It_l + \varsigma_{at} + (P_t * \omega_t) - \Psi + (\delta_t) - \delta_{rt} - (E_t * P_t) - 2(P_t * g_{t+1}) + \lambda_t + (r_{dt} * P_t) + (P_{vt} * F_{vt}) + (P_{pt} * F_{pt}) + U_{tt} + U_{st} + S_{ot} + (r_{rt} * Pt) - (S_{rt})$$

Hence, for the household to prefer ownership to renting, the following conditions must exist:

1. $R_t > \alpha$
2. $\partial R_t / \partial \alpha > 1$; $\partial^2 R_t / \partial \alpha^2 > 0$
3. $\partial S_{rt} / \partial S_{ot} > 1$; $\partial^2 S_{rt} / \partial S_{ot}^2 > 0$
4. $\partial r_{rt} / \partial r_{dt} > 1$; $\partial^2 r_{rt} / \partial r_{dt}^2 > 0$
5. $\partial R_t / \partial P_t \in (0,1)$; $\partial^2 R_t / \partial P_t^2 \in (0,1)$
6. $\{P_t * (g_{t+1} + E_t)\} > \phi_t$
7. $\partial \Omega_d / \partial t \in (-\infty, 1)$; $\partial^2 \Omega_d / \partial t^2 < 0$. This implies that the housing unit must be of a certain minimum quality, before occupancy.
8. $\partial U_{tt} / \partial U_{st} < 0$; $\partial^2 U_{tt} / \partial U_{st}^2 > 0$. This is more of a neighborhood/global condition.

Proposition 5: Stability Bias (Stability of wealth): real estate has traditionally been viewed as being more stable (less volatile prices) than most asset classes. Such stability of wealth provides a certain psychological comfort to home-owners, and hence provides incentives/motivation to purchase homes. Hence, housing units that offer the most expected value 'stability' tend to be preferred. During 1995-2003, the Stability Bias was partly attributable to losses and high volatility experienced in US stock markets, low yields from fixed income investments, declining commodity prices and substantial differences in knowledge of individual investors and institutional investors. Furthermore, price stability may have been inferred by buyers and sellers from the large volume of refinancings of residential mortgages that occurred in the US during 1995-2003 – such refinancings facilitated prices increases by making more capital available per dollar of income reserved for debt service and maintainance. Since the refinancings did not increase homeowners' mortgage payment cash burden (cost of ownership), and home owners can essentially support future mortgage payments from proceeds of said refinancings, home prices could be expected to remain stable and mortgage delinquencies could be expected to remain low in the future - these two classes of expectations were sufficient to cause substantial increases in home prices, and willingness to pay historically high prices for housing units.

Proposition 6 – Price Justification Effect. Price-justification is manifested when there are significant and rapid changes in housing prices that diverge sharply from historical patterns, and are un-related to market fundamentals. See: Stevenson (2004); Gallimore (2004); Hansz (2004). When such changes occur, market participants will typically seek to justify and internalize new prices, availability of loans at such price levels, and directions of prices. The availability of capital due to lower interest rates (which made capital available to buy housing at higher prices during 1995-2003) cannot completely account for the rapid and substantial increases in housing prices in the US during 1995-2003, because: a) housing prices could have still declined despite LTV rates and sale prices previously justified by low interest rates; and b) prospective homebuyers could have simply refused to buy at historically high prices, which were increasing at a much faster rate than household income. Hence, there must have been other psychological factors that contributed to such price justification some of which include:

- Concentration of wealth among an aging population (older than 45 years) that can afford higher home prices; and believe that others will be able to afford future high home prices.
- Tenure Bias – those buying at historically high prices intend to live in such houses for long periods.
- Opportunism.
- Stability Bias

Price Justification is rooted in:

- Tendency of humans to adjust to new conditions; coping.
- Inability of buyer and or broker to control the situation and local/regional housing markets – their transaction is typically just one of many in the market.
- Tendency of buyers to follow trends.
- Reflection Effect – believing that people in the market will think alike.
- Reliance on appraisals.
- Reliance on real estate brokers – who are motivated to obtain the highest sales prices.
- Expectations about property tax rates.
- Opportunism. Expected continued price appreciation.

Proposition 7: Prior Knowledge Bias – In the real estate sector, appraisers are trained in, and use the same valuation formulas and methods. Many investors come from acquisitions, property management or brokerage aspects of the real estate industry, and learnt and used the same or similar due diligence methods and valuation methods. Hence, the rapid price increases that occurred during 1995-2003 could be attributed to Prior Knowledge Bias among appraisers and investors – that is, knowledge of how the opposing party (buyer or seller) would value the property, affects the individual's/investor's propensity to sell or buy a housing unit; and their reactions to price offers.

Proposition 8: Expectations Bias – during 1995-2003, expectations heavily influenced Buyer intent as evidenced by: a) the substantial divergence between the replacement costs of housing units and the market values of housing units, and b) the substantial differences between the growth rates of household incomes and house prices, and c) the relatively low inflation rates during 1995-2003. The expectations were about performance of other asset classes, availability of future buyers, the location and national economies, interest rates, real estate returns, etc.. Meen and Meen (2003); Garvill, Garling, Lindberg and Montgomery (1992). Raess and Von Ungern-Sternberg (2002). Ambrose (2000).

Proposition 9: Willingness To Accept Losses (WTAL) – investors' and individuals' low WTAL may also have contributed to their Propensity-To-Buy in the housing market because of the common perception of real estate and homes as investments with relatively stable prices. Low WTAL for the 1995-2003 period may be reasonably inferred from the: a) declines in the US stock markets, b) unemployment rates, c) localized 'recessions', d) low yields from fixed income securities, e) low dividends, and f) high consumer debt levels that existed in the US from 1997-2003, g) WTAL is typically manifested in buyers' assessments of whether to purchase a housing unit, based on the potential income and or the potential price appreciation, and whether or not the property is subject to rent regulation or rent controls (See: Raess and Von Ungern-Sternberg (2002)), h) the movement of capital between the primary and secondary mortgage markets - many lenders don't hold residential mortgages, but sell them almost immediately after origination – this reflects their low WTAL; i) buyers' psychological perception of expected low volatility of, and or continued increases of real estate values, income potential, and re-sale potential, all within the context of utility gained from use/ownership/lease of the housing unit. See: Ambrose (2000); Schein (2002); Nwogugu (2005c); Bennett, Peach and Peristiani (1998); Sibly (2002); Butler (2000); Merlo and Ortalo-Magne (2004); Ong and Lim (1999); Hort (2000); Krystalogianni and Tsocalos (2004); Meier, Kirchler and Hubert (1999). Investors' and current/prospective home-owners' WTAL was also manifested by the following:

- People moved money out of stock markets (which were declining) and invested in real estate, which was more stable/profitable,
- Percentage changes in the average the ownership tenure and average rental tenure for certain classes of residential property in some regions,
- The reactions of buyers to probabilities of buyers getting mortgage loans,
- Percentage of purchase transactions that involved title insurance,
- Annual changes in buyers' search costs – percentage of buyers that retained brokers, percentage of buyers that had targeted property appraised independently before deciding to buy;
- Annual changes in the percentage of buyers who choose interest-only mortgages as opposed to principal and interest mortgages;
- Annual changes in the percentage of buyers who choose ARMs versus fixed rate mortgages;
- The quarterly percentage changes in the average downpayment (expressed as a fraction of the purchase price) for various classes of housing units;

- The percent of sales/purchase transactions that involved use of escrow accounts;
- The quarterly changes in percentage of buyers who use the internet to research prices and comparable properties before making purchase decisions;
- The quarterly changes in the percentage of a typical buyer's annual income that is committed to housing costs. Schein (2002); Ong and Lim (1999).

The Genesove and Mayer (2001) study is somewhat limited, applies only to a small subset of possible instances/transactions, did not study the impact of loss aversion and or risk aversion on buyers' Propensity-to-buy; and many of the study's results/conclusions were evident and or could be derived from simple correlation analysis. The LOGIT/regression models used in the Genesove and Mayer (2001) 'empirical' study are based on normal distributions and binary choices, both of which don't fit the conditions in the US real estate industry and housing markets - the underlying data (real estate prices and any derivative series) don't follow a normal distribution.

Proposition 10: Risk Shifting Bias - The knowledge that banks can sell mortgages in the secondary markets can create a psychological laxness in the loan-approval process, even though significant portions of the loan approval process are now automated. The laxness occurs when the borrower and the loan agent collude to falsify loan applications to fit underwriting standards, with the knowledge that: a) the loans will be eventually sold off in the secondary market and securitized, and b) that after securitization, any losses from that particular mortgage will be offset by cash flow from other loans in the mortgage pool. Hence, securitization of residential mortgages is conjectured to increase the propensity to commit fraud in the loan application process. The second type of Risk Shifting Bias can occur with the output of the loan approval process data: 1) because most types of loan approval processes are automated, GSEs (government-sponsored enterprises – like FNMA) that want to buy loans may become over-reliant on such automated systems and the associated output which contains false information, 2) although real estate prices and rents varied dramatically by region/city, some research (Ambrose (2002)) has shown that most secondary market loan purchases were done without regard to local economic conditions.

The third type of Risk Shifting Bias is caused by over-reliance on mortgage insurance by the borrower, primary lender and loan purchaser. FHA provides mortgage insurance for a substantial percentage of residential loans in the US; and provides strong incentives for lenders/underwriters to lower mortgage underwriting standards, and for prospective buyers to collude to file false information in order to get mortgages. This affects the quality of underwriting standards programmed into automated underwriting systems and manual processes – generally reducing underwriting quality and shifting risk to FHA.

Proposition 11: Attachments Bias (positive attachment or negative attachment): People can become positively or negatively attached to neighborhoods, a specific property, specific features of a property, certain types of housing units, social circles and certain types of lifestyles. Hort (2000).

This Attachments Bias was evident in the behavior of buyers in the US housing sector during 1995-2003: 1) the substantial use of mortgages for home purchases between 1997-2003, 2) during 1997-2003, most (more than 50%) home sales and purchases by the same household occurred within a relatively narrow time frame (0-7 months) and within the same metro area; 3) the relationship between the length of ownership tenure and "flipper"/non-flipper characteristic during 1995-2003 – the length of ownership-tenure among non-flipper landlords generally increased while that of 'flippers' generally decreased; 4) the increasing tenure of mortgages and changes in mortgage prepayment rates; 5) the dollar amount of capital expenditures used in renovating housing units; 6) the volume of sales of home improvement goods and the volume of home improvements projects undertaken during 1995-2003; 7) the percentage of housing units in previously 'low-income' neighborhoods that were purchased and renovated by residents of such neighborhoods, 8) the percentage of housing units in 'middle-income' and 'high-income' neighborhoods that were purchased by residents of such neighborhoods or by people who are from other neighborhoods but have the same income levels, social capital and social standing. Meen and Meen (2003). Garvill, Garling and Lindberg and Montgomery (1992).

Proposition 12: Mortgage Bias: To some people, Mortgages connote: 1) a certain amount of individual responsibility; 2) a reason/justification for employers to continue to hire them; and 3) a chance to build a credit history; 4) an opportunity to build home equity and savings for future expenses; 5) tax advantages, 6) an opportunity to use leverage to control property that otherwise would be unaffordable. 7) an opportunity to achieve long-held aspirations of home-owenrship, and or wealth, 8) an opportunity to climb the social ladder, and or to move into new and better neighborhoods, 9) an opportunity to diversify household wealth. These factors (and others) are elements of a *Mortgage Bias*. Bennett, Peach and Peristiani (1998); Madsen and MCAleer (2001); Janssen and Jager (2000). In the US, Mexico and Canada, Mortgage-Bias was evidenced by several trends that occurred during 1995-2003: 1) many households who could purchase housing units using only cash, still used mortgages for their purchases; 2) many people who could afford higher down-payments for mortgages, sought and choose low-downpayment or no-downpayment mortgage loans; 3) the volume of no-documentation loans, low documentation loans, and negative-amortization loans increased between 1995 and 2003, 5) the percentage of home purchases financed with mortgage loans increased during 1995-2003; 6) the dollar amount of tax deductions taken by individual tax payers for mortgage payments increased during 1995-2003, 7) the percentage of households with married working couples that had mortgages increased during 1995-2003, 8) the average-age and median-age of mortgage borrowers declined during 1995-2003, 9) the average tenure of a home mortgage (before refinancing or default) increased during 1995-2003, 10) several federal and state government agencies made it policy to increase homeownership, and the primary vehicle was by mortgages, 11) in many cities/towns and neighborhoods where mortgage-financed home-ownership increased substantially, crime rates declined and quality of life improved – many boroughs in New York City are good examples. Courchane, Surette and Zorn (2004). Furthermore, mortgages change the balance of intra-household negotiations, capital allocation, bargaining power, buying habits, classification of disposable income, savings habits, priorization, and other factors. The increased volume of mortgage fraud during 1995-2003 in US, Mexico and Canada can be partly attributed to Mortgage Bias.

Proposition 13: Property Type Bias: People tend to buy and sell properties with which they are most familiar with, or are attached to. Most people who sold their homes in 1995-2003 either used the funds to buy primary/secondary homes, or purchased investment residential properties, even where it was obvious that other forms of real estate (retail, industrial, office) had lower risk (more credit-worthy tenants and less rollover, lower maintainance costs, lower capex) and potentially higher returns. This Property-Type Bias are attributable to knowledge effects, learning curve effects, aspiration levels, need for human interaction, expectations (about returns, etc.), etc.. Garvill, Garling and Lindberg and Montgomery (1992); Janssen and Jager (2000). The Property Type Bias was a causal factor for the subprime loan problems in the US, because the sub-prime loan problems are limited to the US housing industry – mostly single family homes. Similarly, lenders allocated a substantial portion of their capital to financing housing units, instead of other types of real estate, in which they could have earned higher returns and would have been exposed to much less risk (corporate tenants with better credit, cheaper credit enhancement, etc.).

Proposition 14: Prioritization Bias: The decision to purchase a home instead of renting, involves some prioritization process in which individuals and households consciously and sub-consciously order their preferences within the context of available resources and alternatives. The purchase process involves many trade-offs and sometime sacrifices. The magnitude and duration of the prioritization depends on the Buyer's income, state of mind, aspirations, reactions to environments, preferences for specific features/functions of buildings, overall wealth, access to capital, characteristics of the household, loss aversion, Willingness-To-Accept-Losses, and other factors. Hence, individuals' propensity-to-buy will be affected by the nature and duration of the Prioritization required before and after the purchase transaction. Similarly, the decision to make a loan also involves some prioritization processes, in which the lender orders its preferences, selects its risk profile, selects levels of allowable losses and expected returns. Prioritization involves some transaction costs and knowledge acquisition/processing costs (time, etc.). Prioritization Bias contributed to the sub-prime loan problems in the US: a) federal agencies and state agencies made home ownership a major priority during 1995-2003, and created and or indirectly supported many programs (lending and mortgage insurance) that facilitated issuance of sub-prime loans, b) for many individuals with bad credit or insufficient income for mortgage downpayments and mortgage monthly payments, home ownership and maximization of use of current income were priorities and the solution was obtaining sub-prime mortgages.

Proposition 15: Lowest Regret – real estate has one of the lowest *Regret* of all asset classes; where *Regret* is re-defined as: a) Disutility and negative emotions arising from the owner's/holder's inability to modify an asset (controlled in some way – by ownership, lease, option, etc.) or make other changes to the asset, which would create value in the present time, or retrospectively, and b) negative emotions and Disutility arising from loss of the property due to inability to make financing payments (mortgages, taxes) and or to maintain the property. Real estate buyers can make improvements, change the capital structure (and hence the investment-return profiles) of the property, obtain second mortgages, lease the property or make other changes that increase their utility and lower their Regret from ownership. This degree of flexibility is not feasible with publicly traded common stock, commodities, bonds, etc.. Van Dijk, Zeeleenberg and Van Der Pligt (2003). Humprey (2003). Akhter (2003). The decision to purchase a housing unit can be construed as a regret-minimization action. Regret also caused the mortgage fraud sub-prime loan problems that have occurred and continue to occur in the US – a) the complex mortgage loans offered to unqualified borrowers (based on income and or assets) promised, or appeared to have low Regret, b) the Regret-minimizing effect of real estate (described above) increased consumers' impulse to buy real estate, instead of investing in other asset classes, c) the Regret-minimizing effect of real estate also increased the propensity to use mortgage loans in property acquisitions – a relatively low downpayment, or no downpayment enabled buyers to control substantial assets. Regret also caused the mortgage fraud and predatory lending problems in the US: a) the complex mortgage loans offered to unqualified borrowers (based on income and or assets) promised, or appeared to have low Regret – 1) the borrowers were getting an asset that they were not qualified to purchase under normal circumstances, 2) the borrowers had little capital at risk (low down payments or no downpayments), but were not subject to any recourse by lender in case of default, 3) the borrowers were often made to believe that they could modify the mortgages by refinancing them, 4) the borrowers were made to believe that either they were not subject to any civil or criminal penalties, or that the risk of enforcement and prosecution for fraud, was low, or that the risk of detection of mortgage fraud was low due to automation of underwriting processes, and the prohibitive cost of detailed due diligence and the sheer volume of residential loans.

Proposition 16: Occupancy-Stress Bias: Research has proved that there is a continuum of psychological stress arising from home occupancy. Cairney and Boyle (2004). It is conjectured here that this continuum has at one extreme, month-to-month rental, and has debt-financed home ownership in the middle and has equity-financed homeownership at the other end of the spectrum. The growth of home equity (by mortgage reduction or capital appreciation) is a significant reduction of Occupancy Stress. Hence, people/household's choices and decisions pertaining to home purchases versus rental, are influenced by their willingness to accept certain levels of Occupancy Stress. The mortgage foreclosure process is more difficult and expensive than the tenant eviction process. Hence, the increase in the volume of mortgage fraud and sub-prime loans can be attributed to individuals' and household's attempts to reduce their Occupancy Stress.

Proposition 17: Environmental Psychology Effects– Ioannides and Zabel (2003); Meen and Meen (2003); Boyle and Kiel (2001); Militino, Ugarte and Garcia-Reinaldos (_______). Buyer motivation, and many purchases of housing units could be attributed to environmental psychology issues such as:

- Affinity to areas that were undergoing urban renewal – many of the rapid price increases were in these areas; and many middle-income households moved into areas that were deemed 'low-income'. Predatory lending and sub-prime lending have been shown to be often concentrated in certain cities/towns.
- Affinity to social networks (condos, coops, gated communities, etc.). Predatory lending and sub-prime lending typically occurred within specific social strata (groups in the society).
- Proximity to certain areas/industries (eg. Manhattan).
- Reduction in commuting time.
- Re-location to low-density areas
- Noise; traffic; availability parks; street lights; aesthetics of physical structures and roads; etc.

Hence, buyers will be more likely to purchase a home, where environmental psychology factors are deemed appropriate or ideal.

Proposition 18: Rules Effect: Buyers, mortgage brokers and lenders often form conscious and sub-conscious formal/informal rules about the property purchase/sales process. Such rules often arise from their perceptions of the economy, personal preferences, aspirations, value of time, WTAL, knowledge, education, family structure, stated procedures, mortgage laws, bankruptcy processes, foreclosure processes, fraud laws, competition in the industry, trends of property prices, etc.. These rules are vital in establishing prices, making offers, searching for information about housing, processing information, assigning values to attributes of houses and neighborhoods, changes in reservation prices and time-on-the-market. Maxwell (2003). The Rules Effect was a major contributing factor to the sub-prime loan problems and the mortgage fraud problems in the US housing industry: a) buyers essentially internalized improper formal and informal Rules about mortgage qualification, credit risk, downpayments, monthly mortgage payments, penalties for default, penalties for mortgage fraud, etc., b) mortgage brokers created and internalized improper formal and informal rules about credit risk, credit scores, risk mitigation, loan sales, compliance with underwriting standards or major lenders like FNMA, etc., c) lenders also created and internalized improper formal and informal rules about risk, borrower qualification, penalties for non-compliance, etc..

Proposition 20: Income Bias: People's decisions to purchase homes are sometimes influenced by their need for, or their ability to defer certain-income and disposable-income, which in turn, is affected by knowledge, wealth, expectations, expected changes in disposable income, leisure activities, household structure, etc.. Pingle and Mitchell (2002). Karlsson, Garling and Selart (1999); Madsen and MCAleer (2001); Krystalogianni and Tsocalos (2004); Gallimore (2004). A high Income Bias would mean that the individual is less likely to use a mortgage to acquire a housing unit. Hence, the subprime loan problems and mortgage fraud problems that continue to occur in the US can be attributed to very low Income Bias among consumers. Similarly, the lender's propensity to allocate capital to residential mortgage loans depends on its Income Bias. The subprime-loan problems and predatory-lending problems in the US can be attributed to lender's low Income Bias – willingness to defer certain-income by lowering underwriting standards, lowering the mortgage interest rates for the first few years of the mortgage loan, and issuing negative amortization loans.

Proposition 21: Conformity Effect: The Buyer may decide to purchase a home primarily in order to conform to social, economic, psychological and temporal expectations and or traditions, and or peer pressure. Such expectations and or traditions depend on the buyer's age, profession, education, community, heritage, social circles, perceived wealth, knowledge, etc.. Conformity is conjectured to increase with knowledge and education. Bardsley and Sausgruber (2005). Furthermore, the final property sales prices in any community in the US (and sometime the real estate sale/purchase negotiation processes as represented by the listing price, mortgage and final sale price) are essentially public goods or quasi-public goods – they provide information that is critical to market definition. Research has shown that people tend to 'contribute' more to a public good, the more others contribute. In this instance, the 'contribution' is in the form of conformity, confirmation of other people's expectations about price appreciation, and confirmation about validity of a form of financing (mortgage), etc..

Proposition 22: Size Effects: Size apparently does matter in buyers' and sellers' decisions to buy/sell a housing unit. This is manifested in: 1) the percentage of homeowners who 'trade-up' to larger homes, 2) time-on-the-market for houses of various sizes in the same class of residential property, 3) the spread between the listing price and actual sales prices of small homes and large homes in the same class of housing units. Olsen, Donaldson and Pereira (2003); Watson (2003). The sub-prime loan problems can also be partly attributed to Size Effects: a) most borrowers in the sub-prime loan market were trading up from apartments to larger single-family homes, b) most consumers that participated in mortgage fraud did so in order to be able to trade up to larger and more expensive housing units (which they would not ordinarily qualify for).

Proposition 23: Rapid-Profits Bias: Some buyers are motivated or de-motivated to buy, by the anticipation of rapid profits from sales of housing-units, where the holding period is relatively short (1-10 months). The magnitude of the Rapid Profits Bias depends on the buyer's knowledge, existing portfolio, expected ease of obtaining mortgages, interest rates, investment objectives, opportunity cost, tax position, existing liabilities, utility to be gained, etc.. Similarly, lenders were motivated to issue sub-prime mortgages because of the anticipation of rapid profits from loans sales and securitization in secondary mortgage markets. Mortgage brokers and investors had strong incentives to participate in mortgage fraud because of anticipation of rapid profits from greater loan volumes combined with perceived low detection risk, perceived high due diligence costs, and perceived high enforcement costs. The Rapid Profits Bias partly accounts for: 1) the activities of 'flippers' in the US housing markets, 2) the rapid increases in home prices in some regions (sometimes greater than a compounded annual rate of 30%); 3) the shift of capital from stock markets and bond market to residential real estate markets; 4) the increase in the volume of sales/purchase transactions in neighborhoods previously deemed as 'low income' and 'high-crime' areas. Schein (2002); Van Poucke and Buelens (2002); Huck (1999). The Rapid-Profits Bias is relevant because its somewhat contrary to typical and wide-spread notions of residential real estate as a stable long-term investment with modest real capital appreciation.

Proposition 24: Realization Bias: The act of buying a home or residential property is the realization of certain expectations (psychological, temporal, procedural and financial). Hence, buyer's consent to purchases will be more likely, if the results of such actions will be a realization of their expectations - about physical structure, neighborhood, social circles, form of payment, sales price, maintainance costs, timing, etc.. Huck (1999).

Proposition 25: Materialism Effect: The buyer's decision to purchase a home is sometimes determined by his/her materialism. More materialistic buyers are conjectured to be more prone to:

- Buy homes to satisfy egos,
- Buy homes in order to improve their social standing,
- Consider price appreciation and time in their home buying decisions.

Materialism is also a major determinant of the quantity and timing of housing consumption – more materialistic persons will seek to trade upwards even in declining markets and or periods of high interest rates and LTV rates. Materialism is sometimes intertwined with people's perceptions of how others perceive them, and their social/economic aspirations. Watson (2003).

Buyer's Propensity-to-Buy

See: Coloumb (2001); Solan and Vohra (2001); Kobberling and Wakker (2003); Quint and Wako (2004); Sudhoter and Peleg ((2004); Milman (2002); Ghiradato and Marinacci (2001); Renault (2001); Van Poucke and Buelens (2002); Stevenson (2004); Meen and Meen (2003); Levy and Lee (2004). Janssen and Jager (2000). Case, Quigley and Shiller (2005);

Anglin (____); Bennett, Peach and Peristiani (1998); Wu and Colwell (1986); Yavas, Miceli and Sirmans (2001); Andrew and Meen (2003). Hence, the Buyer's Propensity-to-Buy ("PTB") can be calculated as follows:

Let:

C = Carrying costs of ownership – such as mortgage payments, property taxes, maintainance, etc.

G = Gains/profits from ownership

D = Debt secured by property

S = Estimated purchase price or buyer's reservation purchase price

R_s = Expected reinvestment rate

T = capital gains Tax rate

V = Post-purchase property value

J = buyer's family structure. At one extreme, there is the single individual with no kids (0), and at the other extreme, there is the typically family with middle-aged parent, and in the middle, there is senior-citizen couple whose kids have grown up. $J \in (0,1)$. $J \to 1$, as the buyer's family tends towards the typical family with middle-aged parents.

W = levels of additional wealth. This factor represents the buyer's additional of wealth other than regular income from jobs. W could be from trusts, inheritances, trading in securities accounts, lotteries, etc.. $W \in (0,1)$. $W \to 1$, as the purchaser's non-job wealth increases.

E = environmental psychology issues. $E \in (0,1)$. $E \to 1$, as the current and future environmental psychology indicators (crime, open spaces, parks, visual cues, low traffic, etc.) become more positive/favorable/promising.

B = broker influence. $B \in (0,1)$. $B \to 1$, as the real estate Broker's influence (ability to convince purchaser and or seller to take specific actions and or buy) increases.

I = internet effect. $I \in (0,1)$. $I \to 1$, as the purchaser is more familiar with, and uses the Internet more to get property/neighborhood/price information.

K = availability of credit/loans. $F \in (0,1)$. $F \to 1$, as the purchaser ease of obtaining loans increases.

H = household income. $H \in (0,1)$. $H \to 1$, as the property buyer's household income increases.

A = Buyer's age. $A \in (0,1)$. $A \to 1$, as the buyer's age increases.

T = type of property. This ranges from apartments at one end of the spectrum, to townhouses in the middle, and single-family homes at the opposite end of the spectrum. $T \in (0,1)$. $T \to 1$, as subject property becomes more likely to be a single-family home.

L = level of buyer's education. $L \in (0,1)$. $L \to 1$, as the property buyer's education increases (a function of number of academic degrees earned, # of certifications earned, and number of years of university education).

M = required minimum down-payment. Most home purchases require some down-payment that ranges between 2% to 30% of the appraised value of the property. $M \in (0,1)$. $M \to 1$, as the required down-payment and closing costs tends to zero.

O = opportunity cost of down-payment.

α = the annual cost-of-ownership.

R_t = annual Cost-of-Rental

The Buyer's Propensity-to-Buy ("PTB") =

$I_p = [LN\{R_t/\alpha\}]^* [(\partial^2K/\partial H\partial I) + (\partial^2E/\partial A\partial L) + (\partial^2T/\partial H\partial J) + (\partial^2S/\partial B\partial I) + (\partial^2S/\partial W\partial K) + (\partial^2M/\partial S\partial O)]$

$I_p \in (-\infty, +\infty)$

$I_p \to 1$, as Buyer's Propensity-To-Buy increases.

Alternatively,

$I_p = [LN\{R_t/\alpha\}]^* \{[(\partial^2K/\partial H\partial I) / (\partial^2E/\partial A\partial L)] + [(\partial^2T/\partial H\partial J) / (\partial^2S/\partial B\partial I)] + [(\partial^2S/\partial W\partial K) / (\partial^2M/\partial S\partial O)]\}$

CONCLUSION

The implications of the foregoing analysis are that:

- All existing housing demand models and all existing housing price models are inaccurate.
- Social psychology, environmental psychology and household economics/dynamics are major determinants of Buyers' behaviors, sale/purchase processes, housing demand and home prices.
- The Propensity-To-Buy is a relevant element of any market demand equilibrium model or housing price model.

BIBLIOGRAPHY

Adair A and Hutchinson N (2005). The Reporting Of Risk In Real Estate Appraisal Property Risk Scoring. *Journal of Property Investment and Finance*, 23(3): 254-268.

Akhter S (2003). Digital Divide And Purchase Intention: Why Demographic Psychology Matters. *Journal Of Economic Psychology*, 24:321-327.

Ambrose B (2000). Local Economic Risk factors And The Primary And Secondary Mortgage Markets. *Regional Science and Urban Economics*, 30(6): 683-701.

Andrew M and Meen G (2003). Housing Transactions And The Changing Decisions Of Young Households In Britain: The Microeconomic Evidence. *Real Estate Economics*, 31(1):117-138.

Anglin P (1997). Determinants Of Buyer Search In A Housing Market. *Real Estate Economics*, 25(4): 568-578.

Antonides G and Kroft M (2005). Fairness Judgments In Household Decision-Making. *Journal Of Economic Psychology*, 26:902-913.

Baffoe-Bonnie J (1998). The Dynamic Impact Of Macroeconomic Aggregates On Housing Prices And Stock Of Houses: A National And Regional Analysis. *Journal Of Real Estate Finance And Economics*, 17(2); 179-197.

Bardsley N and Sausgruber R (2005). Conformity And Reciprocity In Public Goods Provision. *Journal Of Economic Psychology*, 26:664-681.

Benjamin J and Chinloy P and Jud D (2004). Why Do Households Concentrate Their Wealth In Housing ? *Journal of Real Estate Research*, 26(4):329-343.

Bennett P and Peach R and Peristiani S (1998). Structural Change In The Mortgage Market And The Propensity To Refinance. *Federal Reserve Board Of New York Staff Reports*, 45, September.

Berkowitz J and Hynes R (1999). Bankruptcy Exemptions And The Market For Mortgage Loans. *Journal Of Law and Economics*, 42(2): 809-830.

Boelhouwer P (2002). Capital Accumulation Via Home Ownership: The Case Of Netherlands. *European Journal Of Housing Policy*, 2(2): 167-181.

Boyle M and Kiel K (2001). A Survey Of House Price Hedonic Studies Of The Impact Of Environmental Externalities. *Journal of Real Estate Literature*, 9(2):117-144.

Cariney J and Boyle M (2004). Home Ownership, Mortgage And Psychological Distress. *Housing Studies*, 19(2): 1-10.

Case K and Shiller R (2003). Is There A Bubble In The Housing Market ? Working Paper, Project Muse, http://muse.jhu.edu. *Brookings Papers On Economic Activity*, Vol 2.

Case C, Quigley J and Shiller R (2005). Comparing Wealth Effects: The Stock Market Versus The Housing Market. *Advances In Macro-Economics*, 5(1): 1235-1245.

Coloumb J (2001). Absorbing Games With A Signalling Structure. *Mathematics Of Operations Research,* 26(2):286-303.

Courchane M, Surette B and Zorn P (2004). Subprime Borrowers: Mortgage Transitions and Outcomes. *Journal of Real Estate Finance and Economics*; Special Issue: Subprime Lending - Empirical Studies, 29(4):365-392.

De Bruin A and Flint-Hartle S (2003). A Bounded Rationality Framework For Property Investment Behavior. *Journal of Property Investment and Finance*, 21(3):271-281.

Diaz J, Zhao R and Black R (1999). Does Contingent Reward Reduce Negotiation Anchoring ?. *Journal Of Property Investment and Financing*, 17(4):374-385.

Federal Reserve Bank Of San Francisco (May 2,2002). *House Price Dynamics And The Business Cycle*. 2002-13.

Garvill J, Garling T and Lindberg E and Montgomery H (1992). Economic And Non-Economic Motives For Residential Preferences And Choice. *Journal Of Economic Psychology*, 13(1): 39-59.

Genesove D and Mayer C (2001). Loss Aversion And Seller Behavior: Evidence From The Housing Market. *Quarterly Journal Of Economics*, ______________________.

Ghiradato P and Marinacci M (2001). Risk Ambiguity And The Separation of Utility And Beliefs. *Mathematics Of Operations Research,* 26(4):864-890.

Glaeser E and and Gyourko J (2005). Why Is Manhattan So Expensive ? Regulation And The Rise In House Prices. *Journal Of Law and Economics.*

Green R (2002). Stock Prices And House Prices In California: New Evidence Of A Wealth Effect ? *Regional Science and Urban Economics*, 32(6): 775-783.

Hansz A (2004). Prior Transaction Price Induced Smoothing: Testing And Calibrating The Quan-Quigley Model of The Dis-Aggregate Level. *Journal Of Property Research*, 21(4):321-336.

Himmelberg C., Mayer C and Sinai T (2005). Assessing High House Prices: Bubbles, Fundamentals And Misperceptions. Forthcoming in *Journal Of Economic Perspectives.*

Humprey S (2003). Feedback-Conditional Regret Theory And Testing Regret Aversion In Risky Choice. *Journal Of Economic Psychology*, 25:839-857.

Ihlanfeldt K R (1980). An Inter-temporal Empirical Analysis Of The Renter's Decision To Purchase A Home. *AREUEA Journal*, 8:180-197.

Ioannides Y and Zabel J (2003). Neighborhood Effects And Housing Demand. *Journal Of Applied Econometrics*, 18-563-584.

Janssen M and Jager W (2000). Fashions, Habits And Changing Preferences: Simulation Of Psychological Preferences Affecting Market Dynamics. *Journal Of Economic Psychology*, 22:745-772.

Karlsson N, Garling T and Selart M (1999). Explanations Of Effects Of Prior Income Changes On Buying Decisions. *Journal Of Economic Psychology*, 20:449-463.

Kauko T (2004). Towards Infusing Institutions And Agency Into House Price Analysis. *Urban Studies*, 41(8):1507-1519.

Landis J. and Elmer V. (2002). New Economy Housing Markets: Fast And Furious – But Different ?. *Housing Policy Debate*, 13(2): 233-274.

Levy D and Lee C (2004). The Influence Of Family Members On Housing Purchase And Decisions. *Journal of Property Investment and Finance*, 22(4/5):320-338.

Linneman P (1986). A New Look At The Home Ownership Decision. *Housing Finance Review*, Fall/Winter, 1986.

Malpezzi S (2002). Urban Regulation, The New Economy And Housing Prices. *Housing Policy Debate*, 13(2): 323-333.

Malpezzi S, Englund P, Kim K and Turner B (2004). *Cross Country Models OF Housing Tenure, Rents And Asset Prices: The Effects Of Regulations And Institutions*. Paper prepared For the ENHR Conference, University Of Cambridge, July 2-6, 2004, Cambridge, UK.

Maxwell S (2002). Rule-Based Price Fairness And Its Effect On Willingness To Purchase. *Journal Of Economic Psychology*, 23:191-212.

Meen D and Meen G (2003). Social Behavior As A Basis For Modeling The Urban Housing Market. *Urban Studies*, 40(5/6): 917-918.

Nwogugu M (2005a). Towards Multifactor Models of Decision Making And Risk: A Critique Of Prospect Theory And Related Approaches, Part One and Part Two. *Journal of Risk Finance*, 6(2):150-173.

Nwogugu M (2005b). Further Critique Of Cumulative Prospect Theory And Related Approaches. *Applied Mathematics and Computation* (2006).

Nwogugu M (2005c). *Regret Minimization And Willingness-To-Accept Losses. Applied Mathematics and Computation* (2006).

Olsen J, Donaldson C and Pereira J (2003). The Insensitivity of Willingness To Pay To The Size Of The Good: New Evidence For Healthcare. *Journal Of Economic Psychology*, 25:445-460.

Ong S and Lim S (1999). Risk Mitigation With Buyback Guarantees And Guaranteed Appreciation Plans. *Journal Of Property Investment and Financing*, 18(2):239-253.

Orr A, Dunse N and Martin D (2003). Time On The Market And Commercial Property Prices. *Journal Of Property Investment And Finance*, 21(6): 473-494.

Painter G and Redfearn C ( ). The Role Of Interest Rates In Influencing Long Run Home Ownership Rates. *Journal Of Real Estate Finance And Economics*, 25(2-3): 243-267.

Pingle M and Mitchell M (2002). What Motivates Positional Concerns For Income ? *Journal Of Economic Psychology*, 23:127-148.

Quint T and Wako J (2004). On House-Swapping, The Strict Core, Segmentation And Linear Programming. *Mathematics Of Operations Research,* 29(4):861-877.
Rosen K. (1996). The Economics Of The Apartment Market. *Journal Of Real Estate Research,* 11(3): 215-242.
Schein A (2002). Concern For Fair Prices In The Isreali Housing Market. *Journal Of Economic Psychology*, 23(2): 213-223.
Smith G, Venkataraman M and Dholkia R (1999). Diagnosing The Search Cost Effect: Waiting Time And The Moderating Impact Of Prior Category Knowledge. *Journal Of Economic Psychology*, 20:285-314.
Solan E and Vohra R (2001). Correlated Equilibrium In Quitting Games. *Mathematics Of Operations Research,* 26(3):601-610.
Stevenson S (2004). House Price Diffusion And Inter-Regional And Cross-Border House Price Dynamics. *Journal Of Property Research*, 21(4):301-320.
Van Poucke D and Buelens M (2002). Predicting The Outcome Of A Two Party Price Negotiation: Contribution Of Reservation Price, Aspiration Price And Opening Offer. *Journal Of Economic Psychology*, 23:67-76.
Watson J (2003). The Relationship Of Materialism To Spending Tendencies, Savings And Debt. *Journal Of Economic Psychology*, 24:713-739.
Yavas A, Miceli T and Sirmans C (2001). An Experimental Analysis Of The Impact Of Intermediaries On The Outcome Of Bargaining Games. *Real Estate Economics*, 29(2): 251-276.

Additional Reading

Allen M, Faircloth S and Rutherford R (2005). The Impact of Range Pricing On Marketing Time and Transaction Price: A Better Mousetrap for the Existing Home Market?. *Journal of Real Estate Finance and Economics,* 31(1):71-82.
Arnold M (1999). Search, bargaining and optimal asking prices. *Real Estate Economics*, 27(3):453-481.
Carmon Z and Ariely D (2001). Focusing On The Foregone: How Value Can Appear So Different To Buyers And Sellers. *Journal Of Consumer Research*, 27(3):360-370.
Diehl R, Kornish L and Lynch J (2003). Smart Agents: When Lower Search Costs For Quality Information Increase Price Sensitivity. *Journal Of Consumer Research,* 30(1):56-71.
Harding J, Knight J and Sirmans C (2003). Estimating Bargaining Effects In Hedonic Models: Evidence From The Housing Market. *Real Estate Economics*, 31(4):601-622.
Hendershott P (1997). Equilibrium Models In Real Estate Research. *Journal Of Property Research*, 14:1-13.
Lee B, Chung E and Kim Y (2005). Dwelling Age, Redevelopment, and Housing Prices: The Case of Apartment Complexes in Seoul. *Journal of Real Estate Finance and Economics*, 30(1):55-80.
Lind H (2004). The Story And The Model Done: An Evaluation Of Mathematical Models For Rent Control. Swedish Royal Institute Of Technology, paper #53.

Mays E (1989) A Profit maximizing Model Of Federal Home Loan Bank Behavior. *Journal Of Real Estate Finance and Economics*, 2(4):331-347.

Merlo A and Ortalo-Magne F (2004). Bargaining Over Residential Real Estate: Evidence From England. Working Paper.

Miles D, Sebastian M and Zeldes S (1992). Housing Markets, Consumption and Financial Liberalisation in the Major Economies; Comments. *European Economic Review*, 36(5): 1093-1120.

Panayotis K, Siokis F (2005). Shock and real estate prices in Greece: wealth versus 'credit-price' effect. *Applied Economics Letters*, 12(2):125-128.

Pryce G (2002). Theory and estimation of the mortgage payment protection insurance decision. *Scottish Journal of Political Economy*, 49(2):216-234.

Soman D (2001). The Effects Of Payment Mechanism On Spending Behavior: The Role Of Rehersal And Immediacy Of Payment. *Journal of Consumer Research*, 27(4):460-474.

Wheaton W (1990). Vacancy, Search, and Prices in a Housing Market Matching Model. *The Journal of Political Economy*, 98(6):1270-1293.

Weber E and Nunes J and Hse C (2004). The Effect Of Cost Structure On Consumer Purchase And Payment Intentions. *Journal Of Public Policy and Marketing*, 23(1):43-53.

Yavas A (1992). A Simple Search And Bargaining Model Of Real Estate Markets. *Journal of the American Real Estate and Urban Economics*, 20(4):533-550.

In: Progress in Chaos Complexity Research
Editors: Franco F. Orsucci and Nicoletta Sala
ISBN: 978-1-60456-375-7

Chapter 12

CLIMATIC MEMORY OF 5 ITALIAN DEEP LAKES: SECULAR VARIATION

Elisabetta Carrara[1], Walter Ambrosetti*[1] and Nicoletta Sala[2]
[1]CNR – Istituto per lo studio degli Ecosistemi,
28922 Verbania Pallanza, Italy
[2]Università della Svizzera Italiana (University of Lugano),
6850 Mendrisio, Switzerland

ABSTRACT

The climatic memory of 5 deep lakes (Maggiore, Como, Garda, Iseo and Orta) shows a series of warming and cooling phases from 1887 to 2007 that cannot be attributed to the energy exchanges at the air-water interface alone. This underlines the complexity of the lakes ecosystems' response to the ongoing global change.

Keywords: *Climatic Memory, Complexity, Ecosystems, Global Change.*

The studies on deep water warming of 5 large Italian lakes (Maggiore, Como, Garda, Iseo and Orta), during the period between 1963 and 1999, showed a significant thermal increase in the climatic memory contained in the deep hypolimnion [1] [2]. The analysis performed on the meteorological parameters showed that the thermal energy rise which occurred from 1970 to 1980 was correlated mainly to a decrease in the wind run, while that which occurred from 1987 to 1998 was correlated mainly to the rise in the air temperature compared to that of the surface water, especially during winter [2].

However, the events of 1999 and 2006, caused by particular hydro-meteorological mechanisms such as cold water descent at depth [3], highlighted an ongoing decrease in thermal energy in the climatic memories of all 5 lakes (Figure1). This reduction in thermal

* Email: w.ambrosetti@ise.cnr.it

energy can not be explained only with an analysis of the meteorological parameters responsible for the convective vertical mixing.

Typically after one cooling phase (e.g. 1970 and 1981) a stasis of a few years followed and then a gradual warming of the deep hypolimnetic water [4]. By contrast, following a cooling phase in 1999 and a second one shortly after in 2006, the heat content of the deep hypolimnion of all lakes was drastically reduced (Figure 1); exactly the opposite of what was hypothesized.

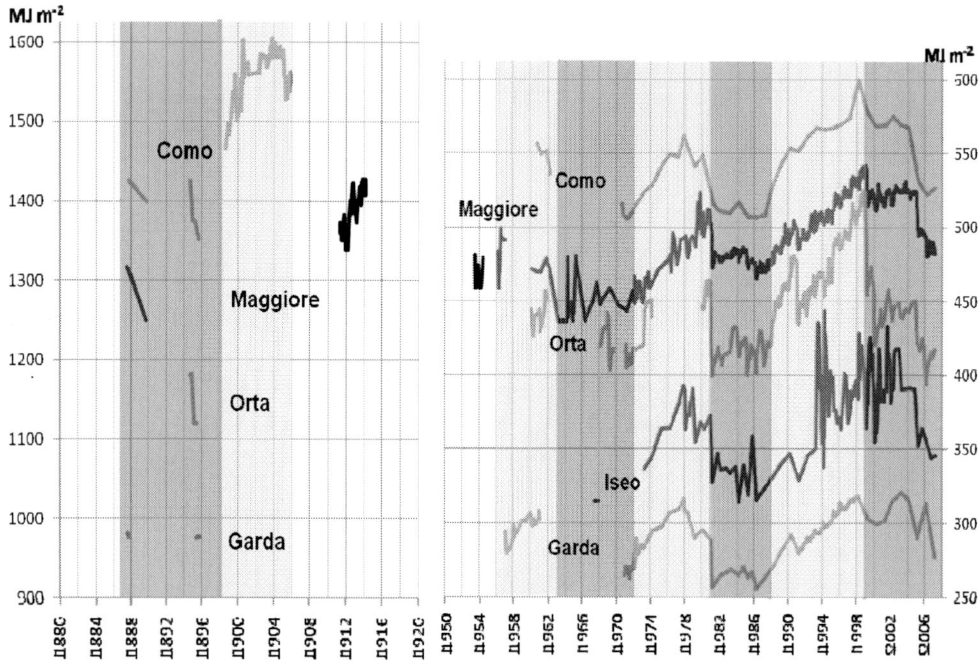

Figure 1. Climatic memory trendlines contained in the hypolimnetic water of 5 subalpine deep lakes from 1887 to 1914 and from 1953 to 2007. Dark and light grey areas indicate 4 cooling and 4 warming phases respectively. The Lake Orta values refer to the right axis.

It was estimated that particular external events could alter only momentarily the rise of the heat content in the climatic memory without really stopping it. It was also envisaged that a return of the climatic memory to the thermal energy levels of the 1950s and 1960s was very unlikely due to the high stability reached by the lake water body (2)(5). Supporting this theory, for example, were the events of 1981 and of 1999 which, although hydrodynamically different, caused a reduction of the heat content of the climatic memory of 5.4% and of 3.9% respectively, halting only momentarily the rise. However, if we sum the reduction of the heat content in 1999 and in 2006, the result is greater than 8%. In fact, Figure 1 clearly shows that the heat content of the climatic memory of all lakes in 2007 is on the same level as the 1950-60 period.

Moving from the more recent data to those pre-1950, dating back to the late 1800s, which although relatively limited in number (Figure 1), are considered adequate to outline the thermal energy situation in the late 19th century and to interpret the evolution of the heat content in the deep hypolimnion [5]. The heat content of the climatic memory in Lake Como from 1898 to 1905 is similar to that of more recent years (1998); meanwhile that of Lake

Maggiore from 1911 to 1914 is higher than 1400 MJ m^{-2}, which is just below the maximum of that of 1979 and 1998. Therefore, it is possible to deduce a warm phase for the deep water of all 5 lakes during the first 20 years of the 20th century.

On the contrary, before 1900 it is possible that a cold phase existed for approximately 15 years. In fact, the climatic memory in Lake Como, measured during two samplings in 1895 shows values below 1400 MJ m^{-2} that have not been measured again since 1950. This is similar to the situation of Lake Maggiore and Orta, while the climatic memory in Lake Garda shows values equivalent to those of more recent years. Evan disregarding the lack of measurements between the two World Wars (1915-50), it is evident that at least 4 warming and 4 cooling phases (see Figure 1 light and dark grey areas) are present in the climatic memory trend.

The results of this research show the importance of particular hydro-meteorological mechanisms in resetting the climatic memory and therefore that its evolution in time does not solely reflect the energy exchanges at the air-water interface [6]. This may also be seen as a form of a "self defense mechanism" underlying the complexity of the lakes ecosystems' response to the ongoing global change.

REFERENCES

[1] Ambrosetti W., and L. Barbanti. *J. Limnol.* 58 (1): 1-9. (1999).

[2] Ambrosetti W., N. and Sala. *Chaos Complex. Lett.* 2 (1): 121-123 (2006).

[3] Ambrosetti W., L. Barbanti, E. A. Carrara and A. Rolla. *Commissione Iternazionale per la protezione delle acque italo-svizzere.* In press (2006).

[4] Livingstone D. *Climatic Change.* 57: 205-225 (2003).

[5] Ambrosetti W., L. Barbanti. and N. Sala. *J. Limnol.* 62 (Suppl.): 1-15. (2003).

[6] Carrara E. A., L. Barbanti and W. Ambrosetti. *Atti del Convegno A.I.O.L.-S.I.T.E. Ancona.* In press (2007).

[7] Robert, C. J. Perez-Losada, G. Schladow, R. Richards and C. Goldman. *Climatic Change.* 76: 121-148 (2006).

In: Progress in Chaos Complexity Research
Editors: Franco F. Orsucci and Nicoletta Sala
ISBN: 978-1-60456-375-7

Chapter 13

ETHOS IN EVERYDAY ACTION NOTES FOR A MINDSCAPE OF BIOETHICS

Franco F. Orsucci
University College, London

Les commencements d'une mutation…
Francisco Varela

1. CONTROL

The Economist magazine of May 23rd 2002, was featuring a special section: "People already worry about genetics. They should worry about brain science too". The cover was about the fear of a near-future of *mind control*: "If asked to guess which group of scientists is most likely to be responsible, one day, for overturning *the essential nature of humanity*, most people might suggest geneticists. In fact neurotechnology poses a greater threat, and also a more immediate one. Moreover, it is a challenge that is largely ignored by regulators and the public, who seem unduly obsessed by gruesome fantasies of genetic dystopias." The journalistic emphasis might be criticized from many points of view, for example: who knows what is the *essential nature of humanity*? Anyway, as mind sciences are progressing, there are several new issues on free will and personal responsibility which are worth some reflections. We will start on some issues related to mental health (not so far from mind sciences).

The public attention is now almost entirely focused on genetics, and the spectre of Nazi *eugenics* is present in the current debate on cloning and stem-cell research. This keeps the ethics of genetic technology high on the political agenda, also for the action of lobbies raising the quarrel over the moral status of embryos or asking to differentiate prevention from eugenics. The Nazis were also victimising the mentally ill and the imprisoned in the name of science and crucial issues are also raised, for example, by programs for the prevention of early psychosis. The social costs of psychoses, due to their frequent chronic evolutions towards social and cognitive impairments, are huge. They are estimated € 15-25 billions per year in Europe: 1 million people become unable to work before 30. In the USA, an average patient's family spends about $ 4000 per year in treatments and 798 hours in assistance (EPOS EU Project).

Though prevention is a crucial issue, the bioethical implications of early psychosis prevention are very complex, as it has been discussed in the NIMH Workshop on Informed Consent In Early Psychosis Research (Heinssen et al. 2001). The potential risks and benefits of early identification in the putatively prodromal phase of schizophrenia were resumed in the following table:

Identification risks	Identification benefits
Unnecessary anxiety or dysphoria, or both	Close monitoring of symptoms
Stigmatization or discrimination, or both, by others	Early identification of psychotic disorder
Self-stigmatization	Reduced treatment delay
Avoidance of developmentally appropriate challenges	Reduced risk of hospitalization
	Reduced risk of behaviors that are harmful or stigmatizing, or both (e.g., suicide attempts, violence, strange or bizarre responses)

The bioethical concern about early psychosis prevention is also related to the fact that most of primary prevention for these disorders includes both psychosocial and pharmacological treatments. The pharmacology of psychosis treatment is based on drugs that decrease neurotrophins and plasticity in the Central Nervous System (Alleva et al, 1996), and may produce cognitive impairment or deficit in the still evolving brain of adolescents. Haloperidol treatment, as probably other similar antipsychotic active on D2 receptors, decreases nerve growth factor levels in the hypothalamus of adult mice.

The benefits of prevention has to be balanced with the risk of false positive diagnoses of prodromal, easily confused with states like mild depersonalization or nuances of depression so frequent and physiological in adolescence.

When it comes to the brain, society tends to regard the distinction between treatment and enhancement as essentially meaningless, just about as *cosmetic*. This flexible attitude towards neurotechnology is likely to extend to all sorts of other technologies that affect health and behaviour, both genetic and otherwise (e.g. plastic surgery). Rather than resisting their advent, people are likely to begin clamouring for those that make themselves and their children look like healthier and happier.

"*History teaches that worrying overmuch about technological change rarely stops it.* Those who seek to halt genetics in its tracks may soon learn that lesson anew, as rogue scientists perform experiments in defiance of well-intended bans. But, if society is concerned about the pace and ethics of scientific advance, it should at least form a clearer picture of what is worth worrying about, and why." (The Economist, cit.).

Current metaphors like 'co-managing evolution' or 'acting as God' are just there to highlight some paradoxes of this evident evolutionary jump of bringing into consciousness, economy and free will genomics and neurosciences (Orsucci, 2002).

2. Discipline

Bioethics should be considered more as an *inter-disciplinary field* rather than a discipline in itself. It has entered its third decade as a self-conscious pursuit, related to but distinct from moral philosophy, theology and health law. But bioethics was and is more than an academic field. Bioethics also manifests itself in clinical and public policy roles. Due to widespread uncertainty about the ethics of certain medical and behavioral procedures and the vicissitudes of legal liability, ethics committees and clinical ethicists have become a part of the social landscape.

Bio-ethicists, wishing to apply ethical knowledge to biomedical matter by following discourse or contextual ethics, were coming to look much like *the philosophical equivalent of psychotherapists*, applying their conceptual schemes to actual problematic cases. Crucially, like psychotherapists, clinical bioethicists could draw on a theoretical framework of principles (autonomy, beneficence, non-maleficence, and justice). Like psychotherapists, bioethicists frequently discard *data* to recognise success or failure in their work, nor even much agree about what *success* means in this context. Moreover, the field was by no means settled about the primacy of the principles, with several competing approaches advanced by important thinkers. They arise specifically in societies that are secular and pluralistic, and among people who do not necessarily identify with an exclusive moral authority, or sometimes, following their "free thinking", do not wholly submit to any authority.

Not much analysis on the idea of *consensus* in bioethics had been done, save for a provocative observation by Stephen Toulmin about his experience as a staff member of the US National Commission for the Protection of Human Subjects in the 1970s, the first ethics commission in the USA. Toulmin reported that the commissioners often surprised him with their ability to agree on the response to a problematic case or policy issue. When he approached them individually and asked them to reconstruct their reasoning, he found they had often reached the same conclusion starting from quite different premises.

3. Evolution

During his lecture of September 9, 2000, in Zürich, Jurgen Habermas (2001; 2003) was moving from Kant's theory of justice to defend the right of abstention in crucial questions, as in his main writing on the "Risks of liberal genetics" takes a firm position following his post-metaphysical thinking.

He reminds that Adorno's *Minima Moralia* (1974) starts just: "the *Gaia Scientia* was considered for such a long time as the proper field of philosophy... the doctrine of the right way". But this way, Adorno assumes, is somehow lost because in the meanwhile ethics has regressed to the stage of *sad science*: at best it can produce just dispersed and aphoristic "meditations on offended life". Adorno moral reflections direct the dark light of pessimism on the shadows of our society, but lack the scientific depth to prepare possible solutions.

As long as philosophy has believed in its capacity to provide a general framework for the right and good life, this would have been a universal model. Universal, just as every religion sees in its founder's life an ideal path of good life and salvation. In this perspective, a large part of philosophy has been a religion without gods.

More recently, ethics tends to circumscribe its normative action to justice, applied to rights and dues of community or individuals. The questions on right and good become context dependent and intermingled with interrogations on the identity of the speaker, considered as object-and-subject. In the end, the result is situation ethics: a system of ethics by which acts are judged within their contexts instead of by categorical principles.

Following this post-modern path, psychotherapies risk to surrogate ethics, by orienting lives or simply dispensing consolations. This can be also related to the always latent social attitude to confuse mental disturbance and ethical disorder. This sort of confusion is neither good for ethics, nor for psychotherapies.

Kierkegaard could be considered the first to answer to this post-modern problem by offering the post-metaphysical concept of "being-able-to-be-yourself", though its meaning might be less clear than we expect. We are confronting, nowadays, with a change in the boundaries and the sense of self and non-self, as the by-product of some sort of an *evolutionary jump* for the human species.

Adults are going to consider as producible and mouldable the genetic matter of their children and plan its design. They might exert on their genetic products a power of disposition. The power to penetrate the somatic base of spontaneous reference and ethical freedom of another person, hitherto considered possible just on things, not on persons.

In this way the boundary between human and non-human is going to change. If and when this change would be accomplished, that day children could consider, for example, their parents, as creators of their genome, responsible for the desired or undesired consequences.

This new possible scenario derives from a change in the social distinction between things and persons, subject and object, and generate new questions:

- How does the self-understanding of a genetically programmed person change?
- How does this change the areas of creativity, autonomy, free-will and equality in human relations?
- Can we foresee a "right to a genetic heritage not compromised by artificial operations"?
- What in Kant was part of the *reign of necessity* is annexed by the *reign of disposability.*

This perspective highlights some similarity between ethics and the immune system: both are the guardians of the Self at the social or at the body level. They define the boundaries between self and non-self in different but correlated domains.

Perhaps we are all (the growing numbers that have entered into the sphere of this transition) "*les commencements d'une mutation*".

4. Local Rules

Proximities and distances within social communities can be a key to understand the implications of this epochal change. At the first Artificial Life Workshop (Langton, 1989) Craig Reynolds presented a computer simulation of bird flocking based on 3 simple rules conditioning an otherwise unconstrained set of *boids*, as he called these simulacra. Each *boid* was required to:

1. Maintain a minimum of distance from other objects, including other boids;
2. Match velocities with neighbouring boids;
3. Move toward the perceived centre of mass of boids in the neighbourhood.

These rules are all local, applying to individual boids, and yet their effect is that of flocking dynamics of striking realism and elegance. *Flocking is here an emergent global phenomenon.* It is interesting to reflect on the extent to which the three rules capture the essence of a standard 3-value set [Liberty (1), Equality (2), and Fraternity (3)] basic to some Western societies. It is worth noting that this value set is based just on simple contextual, local, rules about relations, not necessarily in top-down ideal values.

Of special interest is the way in which the component elements emerging during such self-organization are both mutually constraining and mutually sustaining. Each is a vital local part of the global pattern within the bounded space. A different pattern can be engendered within the same space and with the same values, but the significance is distributed into different clusters, whether differently located, of a different size, or of a different number.

Another basic feature of this cited experiment is that computer simulations based on Artificial Life can show how ethics and values can generate some sort of *in vitro* experiments. The history of humanity has shown enough about the risks of *in vivo* experiments in ethics and politics, and this seemed an opportunity to test ethical hypotheses before any "collateral damage".

These experiments highlight the structuring process of value sets in a small world, by applying just local rule: the self-organization of values is the emergent production of local iterated interactions. This kind of contextual organization is more flexible and adaptable to unexpected events than top-down organizations. From such a perspective it is of greater relevance to recognize the process whereby different kinds of contextual circumstances can evoke the emergence of different patterns from the value space.

Another way to look at such patterning is in terms of the "pathways" that may emerge between different value locations. Ethical values may also affect (or be dependent on) each other to a greater or lesser degree, just as a pattern of mountain valleys may severely condition the nature of relationships between otherwise proximate zones.

The cultural historian William Irwin Thompson synthesizes: "Values are not objects, they are relationships. When you overlay one pattern with another, a third pattern emerges, a moiré pattern" (Thompson, 1996). As Gregory Bateson stated: "Destroy the pattern which connects and you destroy all quality" (Bateson, 1979). This means that values, just as *qualia*, are (embedded in) the mesh of relations.

5. Language and Consciousness

The basic structure of ethics, being a structure of reciprocity, is implied in the structure of communication in language. Anyway, this does not simply mean that we could ground ethics by extricating its basic structure from empirically given forms of language use or social communication. Language reveals itself as the matrix and mediation of relations and the basic medium of the conciliation between people i.e. ethics. *In linguistic games people negotiate how to adapt each other their individual forms of life.*

Following this path, we need to supersede both the naïf representationalist approach to language and the idealist generative grammar. Both these approaches are seriously challenged by advancements in the scientific studies on the nonlinear dynamics of language and neuroscience. Basically, in this perspective, ethical thinking should be considered mostly as a subsystem in language, apt to preserve and maintain secure boundaries for the Self. It is a function that is quite similar to the functioning of the immune system, devoted to preserve our biological self by defining what is Self and what is Non-Self, and discerning the semiosis of confusions and intrusions along their fuzzy boundaries. It might not be casual that bioethics works on the edge between body, mind, and society, just in the same area of functioning of the immune system: in the fuzzy and transitional areas between private and public, where you have to accept the private language paradox *à la* Wittgenstein (Kripke, 1982; Wittgenstein, 1967) *just to conjugate the reasons of body and language.*

7. Timing Actions

We might try to transfer the *boid's* experiment within a human context. As it is based on dynamical interactions, and their evolution, time is a crucial coordinate.

St. Augustine's in the Confessions (specifically book II, 1990) writes on the paradoxes of "a past-and-future containing nowness": *distentio animi*. More recently, the pragmatic psychology of William James, in his Principles of Psychology (1967), proposed the definition of *specious present.*

Time in experience is quite different from time as measured by a clock. Time in experience presents itself not only as linear but also as having a *complex texture* (evidence that we are not dealing with a "knife-edge" present) a texture that dominates our existence to an important degree (Varela in Petitot, 1999).

We could define three levels of temporality:

1. A first level proper to temporal objects and events in the world. This level is close to the ordinary notions of temporality in human experience, which grounds the notion of temporality currently used in physics and computation.
2. A second level, which is quickly derived from the first by reduction, the level of acts of consciousness that constitute object-events. This is the "immanent" level, the level based on the "internal time" of acts of consciousness. Their structure forms the main body of the phenomenological analysis in Husserl's Lectures (Husserl, 1980).
3. Finally (and this is the most subtle level of analysis), these first two levels are constituted from another level where no internal-external distinction is possible, and which Husserl calls the "absolute time constituting flow of consciousness".

As phenomenological research itself has repeatedly emphasized, *perception is based in the active interdependence of sensation and movement.* Several traditions in cognitive research have, as well and in their own way, identified the link between perception and action as key. *It is this active side of perception that gives temporality its roots in living itself.* Within this general framework, we will concentrate more precisely on the structural basis and consequences of this *sensory-motor integration* for our understanding of temporality.

This overall approach to cognition is based on *situated embodied agents*. Varela (Varela, 1991; Thompson, 2001) has introduced the name *enactive* to designate this approach more precisely. It comprises two complementary aspects: (I) the ongoing coupling of the cognitive agent, a permanent coping that is fundamentally mediated by sensory-motor activities; (2) the autonomous activities of the agent whose identity is based on emerging, endogenous configurations (or self-organizing patterns) of neuronal activity.

Enaction implies that sensory-motor coupling modulates, but does not determine, an ongoing endogenous activity that it configures into meaningful world items in an unceasing flow.

From an enactive viewpoint, any mental act is characterized by the concurrent participation of several functionally distinct and topographically distributed regions of the brain and their sensory-motor embodiment. From the point of view of the neuroscientist, it is the complex task of relating and integrating these different components that is at the root of temporality.

These various components require *a frame or window of simultaneity that corresponds to the duration of lived present*. There are three scales of duration to understand the temporal horizon just introduced:

- basic or elementary events (the "1/10" scale);
- relaxation timefor large-scale integration (the "1" scale);
- descriptive-narrative assessments (the "10" scale).

The first level is already evident in the so-called *fusion interval* of various sensory systems: the minimum distance needed for two stimuli to be perceived as non-simultaneous, a threshold that varies with each sensory modality. These thresholds can be grounded in the intrinsic cellular rhythms of neuronal discharges, and in the temporal summation capacities of synaptic integration. These events fall within a range of 10 msec (e.g., the rhythms of bursting inter-neurons) to 100 msec (e.g., the duration of an EPSP/IPSP sequence in a cortical pyramidal neuron). These values are the basis for the 1/10 scale.

Behaviorally, these elementary events give rise to *micro-cognitive phenomena* variously studied as perceptual moments, central oscillations, iconic memory, excitability cycles, and subjective time quanta. For instance, under minimum stationary conditions, reaction time or oculomotor behavior displays a multimodal distribution with a 30-40 msec distance between peaks; in average daylight, apparent motion (or "psi-phenomenon") requires 100 msec.

This leads naturally to the second scale, that of long-range integration. Component processes already have a short duration, on the order of 30-100 msec; how can such experimental psychological and neurobiological results be understood at the level of a fully constituted, normal cognitive operation? A long-standing tradition in neuroscience looks at the neuronal bases of cognitive acts (perception-action, memory, motivation, and the like) in terms of cell assemblies or, synonymously, neuronal ensembles. A cell assembly (CA) is a distributed subset of neurons with strong reciprocal connections.

The diagram depicts the three main hypotheses. A cognitive activity (such as head turning) takes place within a relatively incompressible duration, a *cognitive present*. The basis for this emergent behavior is the recruitment of widely distributed neuronal ensembles through increased frequently, coherence in the gamma (30-80 Hz) band. Thus, the corresponding neural correlates of a cognitive act can be depicted as a synchronous neural hypergraph of brain regions undergoing bifurcations of phase transitions between cognitive present contents.

Recently this view has been supported by widespread findings of oscillations and synchronies in the gamma range (30-80 Hz) in neuronal groups during perceptual tasks. Thus, we have neuronal-level constitutive events that have a duration on the 1/10 scale, forming aggregates that manifest as incompressible but complete cognitive acts on the 1 scale.

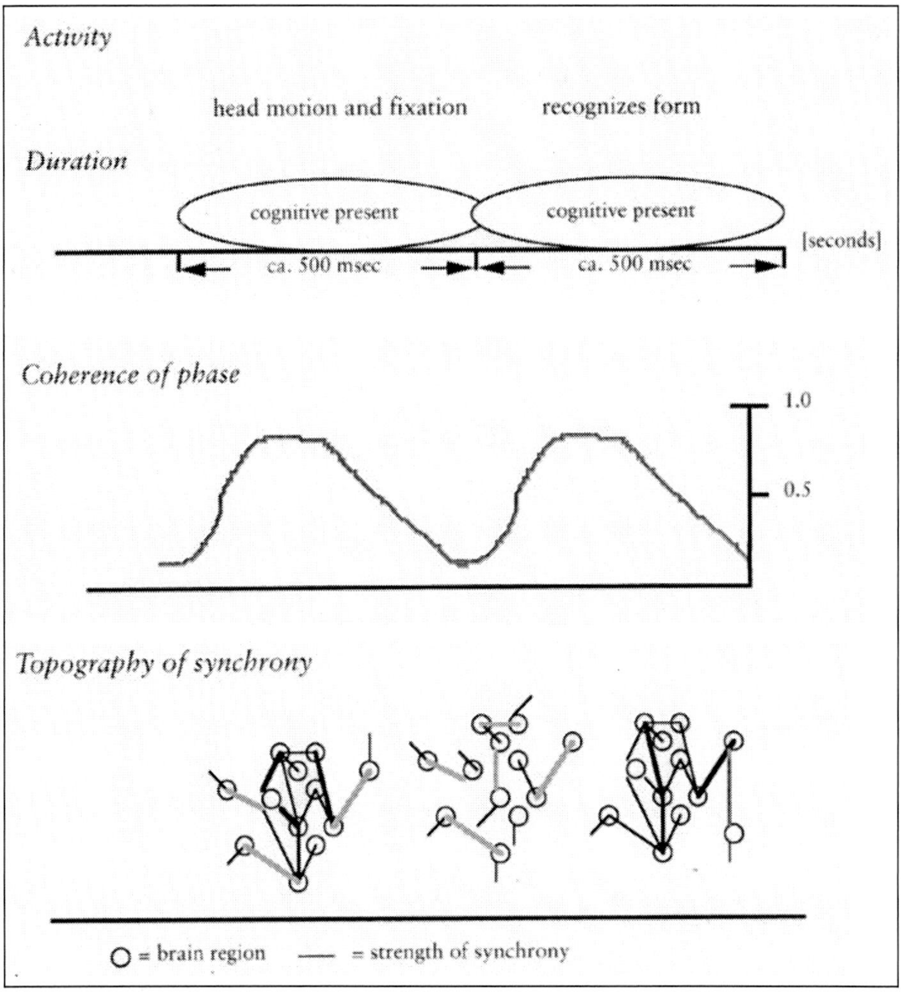

Windows of time (Varela, 1991).

This completion time is dynamically dependent on a number of dispersed assemblies and not on a fixed integration period; in other words it is the basis of the origin of duration without an external or internally ticking clock.

Nowness, in this perspective, *is therefore pre-semantic* in that it does not require a re-memoration in order to emerge. The evidence for this important conclusion comes, again, from many sources. For instance, subjects can estimate durations of up to *2-3 seconds* quite precisely, but their performance decreases considerably for longer times; spontaneous speech in many languages is organized such that utterances last 2-3 seconds; short intentional movements (such as self-initiated arm motions) are embedded within windows of this same duration.

This brings to the fore the third duration, the 10 scale, proper to *descriptive-narrative* assessments. In fact, it is quite evident that these endogenous, dynamic horizons can be, in turn, linked together to form a broader temporal horizon. This temporal scale is inseparable from our descriptive-narrative assessments and linked to our linguistic capacities. It constitutes the "narrative center of gravity" in Dennett's metaphor (Dennett, 1991), *the flow of time related to personal identity*. It is the continuity of the self that breaks down under intoxication or in pathologies such as schizophrenia or Korsakoff's syndrome. As Husserl points out, commenting on similar reasoning in Brentano: "We could not speak of a temporal succession of tones if ... what is earlier would have vanished without a trace and only what is momentarily sensed would be given to our apprehension" To the appearance of the just-now one correlates two modes of understanding and examination (in other words, valid forms of donation in the phenomenological sense):

(I) remembrance or evocative memory and (2) mental imagery and fantasy.

The *Ur-impression* is the proper mode of the now, or in other words, it is where the new appears; impression intends the new. Briefly: *impression is always presentational, while memory or evocation is re-presentational.*

This behavior embodies the important *role of order parameters in dynamical accounts.*

Order parameters can be described under two main aspects:

1. The current state of the oscillators and their coupling, or initial conditions; and
2. The boundary conditions that shape the action at the global level: the contextual setting of the task performed, and the independent modulations arising from the contextual setting where the action occurs (namely, new stimuli or endogenous changes in motivation).

8. Neurodynamics of Time

The neurodynamics of time we have been pursuing is essentially based on nonlinear coupled oscillators. As we saw, this class of dynamical systems finds its wealth of behavior in the fact that constitutional instabilities are the norm and not a nuisance to be avoided (Varela, in Petitot, 1999). The case of multi-stability makes this quite evident experientially: the percepts flip from one to another (depending on order parameters) by the very nature of the geometry of the phase space and the trajectories.

This is a highly schematic diagram of the current view that a complex, chaotic dynamic should be regarded as having geometry in phase space where multiple instabilities are found locally (gray lines). A system's trajectory jumps (in black lines) constantly between local instabilities, in an unceasing flow, under the modulation of boundary conditions and initial conditions. Experimental evidence is given for local saddle instabilities in the time series from the cortex of an implanted epileptic patient. The peaks of the local discharge in the temporo-occipital cortex are followed in their return map or Poincaré section.

"Even the most precise consciousness of which we are capable is affected by itself or given to itself. The very word consciousness has no meaning apart from this duality," (Merleau-Ponty, 1962). Alterity is the primary clue for time's constitution. We are affected not only by representations and immanent affection ("*affection de soi par soi*"), but by the inseparable alterity from the sphere of an *ego-self.*

The very distinction between *auto* and *hetero* ceases to be relevant, since in all cases it all is brought down to the same manifestation: it is a question of "*something other*", the experience of an alterity, a difference in the identity of is constitutive in the paradoxical nature of the shifter called Ego. Subject and object almost at the same timewith an infinitesimal delay.

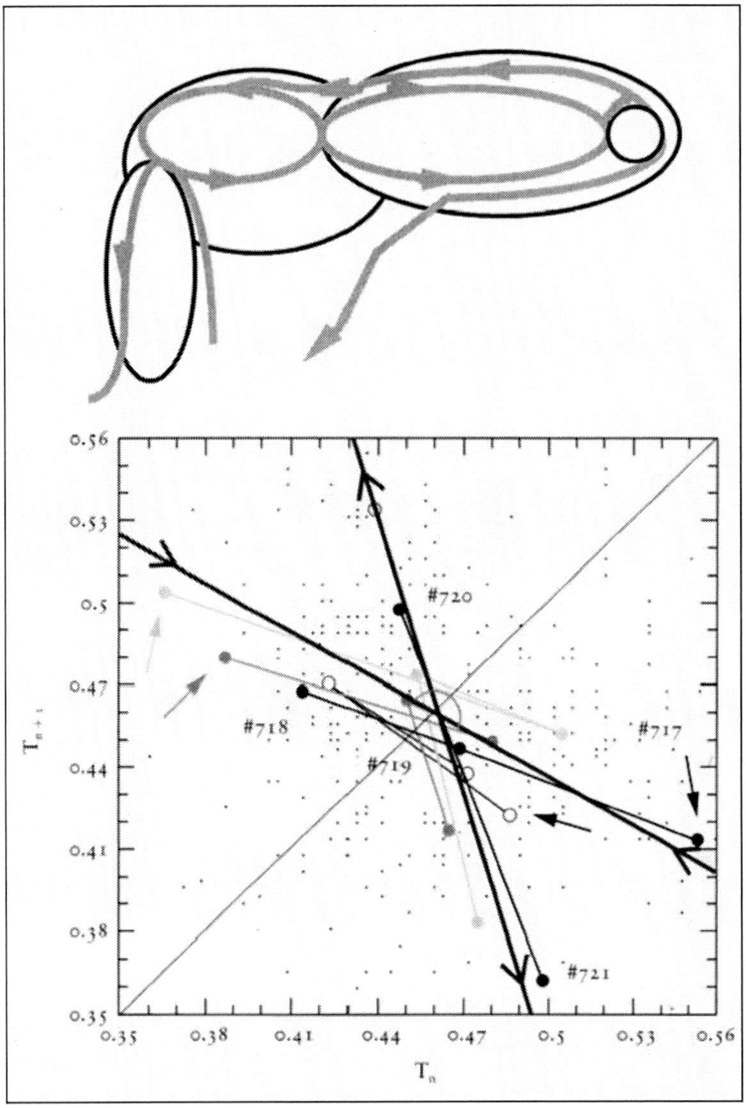

Multiple instabilities (Varela, in Petitot, 1999).

We seek a non-dual synthesis whereby *affect* is constitutive of the self and at the same time contains a radical openness or unexpectedness concerning its occurring. This bootstrap principle seems to be present in a variety of natural systems and has recently been referred to as *operating at the edge of chaos*, or *self-organized criticality*. For example, this idea provides a renewed view of evolution, since it provides an answer to the old nature (genetic expression) versus nurture (environmental coupling conditions) dilemma. In this synthetic view (Kauffman, 1993) the relation between natural forms and the selection process in their ecological embedding is not one of contradiction but precisely mutual imbrications. This built-in shiftiness, enfolding trajectories and geometry, gives a natural system a possibility of always staying close to regions in phase space that have multiple resources (e.g., at least two in bi-stable visual perception).

9. Ethos in Action

The subject of ethics has the ambiguous nature built on the imbrications between the immediate copying of bodies and the reflections of social institutions. These are embedded in the intra subjective dialectics between presentation and representation, intentionality and reflection. That is a result of the repeated meetings of doubles. There is a "readiness-for-action a micro-identity and its corresponding level a micro-world...we embody streams of recurrent micro-world transitions" (Varela, 10). And these micro worlds are nested and imbricated in other worlds. There is a small time-lag somewhere, like a butterfly-wings'-flap, that generates through multiple cascades the doubling and mirroring effect that we call consciousness. This constitution of the subject founds the status of decision, volition and action (see Gazzaniga, 2005).

In presentation and action there is an embedding of representation and imagination (Lakoff and Johnson, 1999). The embodied mind emerges and grows (bottom-up) on the basic reflexive function as a direct parameter in biological processes. The ethical reflection is a meta-cognitive function (top-down): "the overall reflective process can embed more conceptual and linguistic functions in the brain than the reflexive component alone" (Siegel, 2007). Some authors use the terms synonymously but, we prefer to use a different terminology to stress a conceptual and factual difference. Reflexivity will be direct and non-conceptual: it implies an immediate capacity of awareness without effort, or intellectualization. In reflexivity the interface is just like your own skin, and it is useful to remind that the embryological origin of skin, brain and mind is the same. The ectoderm, our primary interface, is the outermost of the three primary germ layers of an embryo and the source of the epidermis, the nervous system, the eyes and ears: i.e. interfaces. Reflexions happen at a very pre-cognitive stage, before any higher order metacognition might be established through a bottom-up spiralling self-organization process. The reverse process is also important as the reflective metacognitive function can impinge on the medium-long term on reflexivity dynamics and neuroplasticity modifications.

The emergence of *ethos in action* is the complex stream generated by the interaction of reflexive and reflective processes.

REFERENCE LIST

Adorno TW (1974) *Minima moralia*, London: New Left Books.

Alleva E, Della Seta D, Cirulli F, Aloe L. Haloperidol treatment decreases nerve growth factor levels in the hypothalamus of adult mice. *Prog Psychopharmacol Biol Psychiatry* 1996;20:483-9.

Aloe L, Iannitelli A, Bersani G, Alleva E, Angelucci F, Maselli P, et al. Haloperidol administration in humans lowers plasma nerve growth factor level: evidence that sedation induces opposite effects to arousal. Neuropsychobiol 1997;36:65-8.

Augustine, Hill, E., Rotelle, J. E., and Augustinian Heritage Institute (1990). *The works of Saint Augustine, a translation for the 21st century*. Brooklyn, NY: New City Press.

Bateson, G. (1979). *Mind and nature a necessary unity*. (1st ed ed.) New York: Dutton. Cazzaniga

Dennett, D. C. (1991). *Consciousness explained*. (1st ed ed.) Boston: Little, Brown and Co.

EPOS, the "European Prediction of Psychosis Study", (EU Contract N° QLG4-CT-2001-01081), http://www.epos5.org/

Gazzaniga, M.S. (2005). *The Ethical Brain*. New York: The Dana Press.

Habermas J (2001) *On the pragmatics of social interaction*, Cambridge, UK:Polity Press.

Habermas J (2003) *The future of human nature*, Cambridge, UK:Polity Press.

Heinssen RK, Perkins DO, Appelbaum PS and Fenton WS (2001) Informed Consent in Early Psychosis Research: National Institute of Mental Health Workshop, November 15, 2000, *Schizophrenia Bulletin*, 27(4):571-584,2001.

Husserl, E. (1980). *Collected works*. Boston: The Hague.

James W (1967) *The writings of William James* (McDermott JJ, editor) New York:Random House.

Kauffman, S. A. (1993). *The origins of order self organization and selection in evolution*. New York: Oxford University Press.

Kripke, S. A. (1982). *Wittgenstein on rules and private language an elementary exposition*. Cambridge, Mass: Harvard University Press.

Lacan, J. (1966). *Écrits*. Paris: Editions du Seuil.

Lacan, J. (1992). *The ethics of psychoanalysis, 1959-1960*. (1st American ed ed.) New York: Norton.

Lakoff, G. and Johnson, M. (1999). *Philosophy in the flesh the embodied mind and its challenge to Western thought*. New York: Basic Books.

Langton, C. G. (1989). *Artificial life, the proceedings of an Interdisciplinary Workshop on the Synthesis and Simulation of Living Systems, held September, 1987, in Los Alamos, New Mexico*. (v. 6 ed.) Redwood City, Calif: Addison-Wesley Pub.

Marcuse, H. (1962). *Eros and civilization a philosophical inquiry into Freud*. New York: Vintage Books.

Merleau-Ponty, M. (1962). *Phenomenology of perception*. New York: Humanities Press.

Orsucci F (2002) *Changing Mind*: transitions in natural and artificial environments. World Scientific, Singapore and London.

Petitot J (editor) (1999) *Naturalizing phenomenology*, Stanford, Calif:Stanford University Press, 1999.

Siegel DJ (2007) *The mindful brain*, New York:Norton.

Thompson, W. I. (1996). *Coming into being, artifacts and texts in the evolution of consciousness*. New York: St. Martin's Press.

Thompson E and Varela FJ (2001) Radical embodiment: neural dynamics and consciousness. *Trends Cogn Sci.* 5 (10):418-425.

The Economist (May 23rd 2002) *People already worry about genetics. They should worry about brain science too,* Special Section.

Varela FJ, Thompson E, and Rosch E. (1991) *The embodied mind*, Cambridge, Mass:MIT Press.

Varela FJ (1999) *Ethical know-how*, Stanford, Calif:Stanford University Press,.

Wittgenstein, L. (1967). *Philosophical investigations*. Oxford.

INDEX

B

C

D

E

F

G

H

I

J

K

L

M

N

O

P

Q

R

S

T

U

V

W